Exploring the Next Frontier

The 1960s and early 1970s saw the evolution of frontier myths even as scholars were renouncing the interpretive value of myths themselves. Works such as Joe Haldeman's *The Forever War* exemplified that rejection using his experiences during the Vietnam War to illustrate the problematic consequences of simple mythic idealism. Simultaneously, Americans were playing with expanded and revised versions of familiar Frontier Myths, though in a contemporary context, through NASA's lunar missions, *Star Trek*, and Gerard K. O'Neill's *The High Frontier*.

This book examines the reasons behind the exclusion of Frontier Myths to the periphery of scholarly discourse and endeavors to build a new model for understanding their enduring significance. This model connects NASA's failed attempts to recycle earlier myths, wholesale, to *Star Trek*'s revision of those myths and rejection of the idea of a frontier paradise to O'Neill's desire to realize such a paradise in Earth's orbit. This new synthesis defies the negative connotations of frontier myths during the 1960s and 1970s and attempts to resuscitate them for relevance in the modern academic context.

Matthew Wilhelm Kapell holds master's degrees in history and anthropology and a PhD in American Studies from Swansea University (UK). He has taught those disciplines as well as film at multiple universities. Previous work includes books on film and digital games and journal articles in African, urban, and utopian studies.

Routledge Advances in American History

1 **The Origin of Organized Crime in America**
The New York City Mafia, 1891–1931
David Critchley

2 **Science, Sexuality, and Race in the United States and Australia, 1780s–1890s**
Gregory D. Smithers

3 **Public Health and the US Military**
A History of the Army Medical Department, 1818–1917
Bobby A. Wintermute

4 **Exploring the Next Frontier**
Vietnam, NASA, *Star Trek* and Utopia in 1960s and 70s American Myth and History
Matthew Wilhelm Kapell

Exploring the Next Frontier

Vietnam, NASA, *Star Trek* and Utopia in 1960s and 1970s American Myth and History

Matthew Wilhelm Kapell

Routledge
Taylor & Francis Group
LONDON AND NEW YORK

First published 2016 by Routledge

2 Park Square, Milton Park, Abingdon, Oxfordshire OX14 4RN
52 Vanderbilt Avenue, New York, NY 10017

Routledge is an imprint of the Taylor & Francis Group, an informa business

First issued in paperback 2019

Library of Congress Cataloging-in-Publication Data
Names: Kapell, Matthew.
Title: Exploring the next frontier : Vietnam, NASA, Star Trek and utopia in 1960s and 70s American myth and history / by Matthew Wilhelm Kapell.
Description: New York, NY : Routledge, 2016. | Series: Routledge advances in American history ; 4 | Includes bibliographical references and index.
Identifiers: LCCN 2015044957 (print) | LCCN 2016001642 (ebook) | ISBN 9781138188570 (alkaline paper) | ISBN 9781315641911 (ebook) | ISBN 9781315641911 ()
Subjects: LCSH: Frontier and pioneer life—United States—Philosophy. | Myth—Political aspects—United States—History—20th century. | Frontier thesis. | United States—Territorial expansion—Philosophy. | United States—History—1961–1969. | United States—History—1969– | Vietnam War, 1961–1975—Influence. | United States. National Aeronautics and Space Administration—History—20th century. | Star trek (Television program) | O'Neill, Gerard K. High frontier.
Classification: LCC E179.5 .K368 2016 (print) | LCC E179.5 (ebook) | DDC 973.01—dc23
LC record available at http://lccn.loc.gov/2015044957

ISBN: 978-1-138-18857-0 (hbk)
ISBN: 978-0-367-26396-6 (pbk)

Typeset in Sabon
by Apex CoVantage, LLC

This book is for
Amy G. Kapell.
It would not have happened
without all the Post-it Notes.

Contents

Acknowledgments

As is always the case with a work of this size the debts are innumerable and cannot be repaid in full. At the University of Michigan–Dearborn, Sharon Lobdell was instrumental in tracking down multiple obscure sources in the United States and getting them to me in the United Kingdom. Nathanial Hoover helped in procuring the 1967 "writer's bible" for *Star Trek*. Now a professor of law at the University of Missouri–Kansas City, but formerly the chief prosecutor during thc United States Arıny's case against some of the perpetrators of the My Lai massacre, William Eckhardt helped confirm quotes from the original courts-martial. At Southern Methodist University's Meadows School of the Arts, Rick Worland was gracious enough to supply access to his previous research on *Star Trek*. At the University of Wisconsin–Milwaukee, John W. Jordan provided similar access to his rhetoric scholarship on NASA and John Kennedy.

Science fiction author and Massachusetts Institute of Technology writing professor Joe Haldeman was overly generous in answering questions about his writings on Vietnam, his novel *The Forever War*, and *Star Trek*. I'm grateful, also, for her permission to quote our e-mail exchanges throughout. Richard Ravalli, historian at William Jessup University, offered insight into the use of both *myth* and *history* here. That is, he found issues of gaps of implications of those terms and deserves thanks for such a task. John Shelton Lawrence, emeritus professor of philosophy at Morningside College in Sioux City, Iowa, read most of this work in draft and offered extensive notes when asked—and just encouragement when I couldn't bear his exactitude. Similar encouragement was provided by Ace G. Pilkington, who, despite the fact that his name implies that he might be a gumshoe private eye, is actually professor of English and history at Dixie State University, Utah. His ability to help navigate between differing academic traditions helped me avoid many a mistake here. Andrea Adams at the University of California Davis talked through aspects of what this work needed to "do" in ways that seem designed to make sure I didn't end up running around, flapping my arms, and shouting gibberish. And, she wrote the blurb. At the library at Woodland Community College in California Traci Johnson was a delight in getting some final sources for this work. Dr. Sarah Trott of Swansea University answered too many questions about institutional and cultural differences between the United States and the United Kingdom. She should have been an anthropologist because she's so good at it.

Zoe B. Sluka-Kapell offered the kind of encouragement and space that could act as a metaphor for topic of space in this work. Her aid can never be acknowledged enough.

In terminology that fits with this work—again, metaphorically—it could easily be said of Professor Stephen McVeigh that "he hits what he aims for." If I hit my targets here it is because he steadied my hand while I aimed and, when necessary, reloaded for me. And, eventually, he both accepted sentences beginning with *And* and the metaphysical significance of the em dash.

All of these debts are unpayable, of course. But of the list of debts that cannot be repaid, the largest belongs to my wife, Amy Kapell. It is only on that debt that I can lay claim to the rest of my life to attempt to pay it back. *Immortalis amor semper.*

And here stands man, stripped of myth, eternally starving, in the midst of all past ages, digging, and scrabbling for roots, even if he must dig for them in the most remote antiquities. What is indicated by the great historical need of unsatisfied modern culture, clutching about for countless other cultures, with its consuming desire for knowledge, if not the loss of myth, the loss of the mythical home, the mythical womb?

Friedrich Nietzsche
The Birth of Tragedy (1872)[1]

Note

1 Friedrich Nietzsche, *The Birth of Tragedy*, trans. Shaun Whiteside (London: Penguin Books, 1993), 109–110.

1 Introduction
1969 and an American Mythos

> [T]he "west"—(How much does that expression mean to include? I never have been able to discover its limits).
>
> —Caroline Kirkland, 1839 (1855, 11)

Walter Cronkite cried.

From behind his newscaster's desk as he watched, with about 500 million other people, as Neil Armstrong descended the ladder on the lunar module *Eagle* and stepped on to the surface of the moon, Cronkite removed his glasses and allowed himself—"the most trusted man in America"—to be overcome with emotion. On the following morning, July 21, 1969, the *New York Times* ran one of the largest headlines it had ever published. In all capital letters it read, simply, "MEN WALK ON MOON."

Times science correspondent John Noble Wilford described the event as

> The realization of centuries of dreams, the fulfillment of a decade of striving, a triumph of modern technology and personal courage, the most dramatic demonstration of what man can do if he applies his mind and resources with single-minded determination . . . The moon, long the symbol of the impossible and the inaccessible, was now within man's reach, the first port of call in this new age of spacefaring.
>
> (1–2)

Even Walter Cronkite compared it to a dream. "Right now after seeing it happen, knowing that it happened," the CBS anchor commented, "it still seems like a dream" (Cronkite 1969). But as a "first port of call," the dream of the moon lasted but a short time. The NASA Apollo missions ended in 1972, and no human has returned since. But the rhetoric of Cronkite and Wilford was neither foreign to Americans nor specific to the *Times* and CBS News. On the same day the *Washington Post* called the event part of a "historic moment in man's quest for his origins" (O'Toole 1969, 1).

But perhaps it was the front-page story in the *Seattle Times* that parsed the rhetoric most carefully. Capturing both the very American part of the dream, and giving it a specific form from a city of the American West, William W. Prochnau declared, "Never again would there be a last frontier" (1969, 1).

Yet the "frontier," as dreamed by the NASA missions to the moon, was less successful than the nearly half-billion people who watched that first moon landing might suggest. American anthropologist Henry A. Selby recalled the event less than two years later. Describing his reaction—and the reaction among those he was living with as an anthropologist in Mexico at the time—Selby admitted that the local Zapotec Indians did not believe any of it was real. As an American in Mexico, with the Zapotec and other American anthropologists as well, Selby described the difference between the two groups' willingness to accept what they were seeing on television. Of the Americans Selby said, "We smiled, because we believed in the moon landing. It was a reality to us, and we shared in Walter Cronkite's tearful awe at the mystery of the knowledge and technology that made it possible. Cronkite was creating modern myth before our eyes" (Wetzels 1973, 161).

This book deals with the rejection of "myth"—and specifically "frontier myths"—in Americanist academics at the end of the Vietnam War. At the same time, ironically, Americans were beginning to experiment with new forms of frontier mythologies. The perceived rejection of frontier myths can be found most easily with a literary history of Joe Haldeman's science fiction novel about Vietnam titled *The Forever War*. The many forms of experimentation of new frontier myths are detailed in the cultural history of NASA's lunar landings, the importance *Star Trek*'s original mission of exploring interstellar frontiers while rejecting paradise, and Princeton University physics professor Gerard K. O'Neill's idea of an orbital frontier utopia supported by technological prowess that he called the High Frontier. In short, this book deals with various ways of constructing "myths" about "frontiers" at the very moment in American culture and academia in which both terms were being rejected.

The irony of examining these topics in this fashion may be obvious to scholars in a number of fields. In American Studies the idea of "myth" as a method of scholarship was being rejected from the discipline at the same time the specific topics of this book were being manifest. The lunar landings may have been a kind of "new symbols of possibility" that Leo Marx called for in 1964, but it was already being rejected as such a symbol when Armstrong, Aldrin, and Collins journeyed to the moon (Marx 1964, 365). In academic history, the 1960s saw the final rejection of any conception of "the frontier" as an organizing principle for understanding U.S. history. Both media studies and literary history the idea of "myths of the frontier" were becoming anathema with the emergence of various schools of European thought filled with its own vocabulary of neologisms designed to make such apparently simple ideas as "myth" seem unimportant. Finally, those studying science, technology, and related areas could find little reason to examine Gerard K. O'Neill's ideas of a High Frontier simply because they never came to fruition—they were, from the perspective of scholars in that field, simply not worthy of study, other than as a failed thought experiment.

So, while the war in Vietnam, NASA's missions, *Star Trek* and O'Neill all present ideas of new mythic frontiers scholars have labored to find other ways to analyze them. The Apollo 11 landing has been examined as the culmination of John Kennedy's New Frontier and as a political outgrowth of the Cold War. But, regardless of how it is presented there can be little doubt that whatever fascination Americans had for lunar exploration in 1969, it was a fascination that quickly faded. And, considering the American culture's long history with the romantic dream and very real myth of "the frontier," this should be surprising. It is all the more surprising because the frontier was a significant presence in 1969 America. Earlier in the year American involvement in South-East Asia had reached in numerical peak of over half a million servicemen and women in the region.[1] As much as NASA's Apollo missions projected all that Americans hoped was positive about their frontier ideology into space, more Americans were serving in the "Indian country" of Vietnam than at any time before. In February Cronkite had reported from the ground in Vietnam, declaring that "we are mired in a stalemate," and the autumn would see the story of the My Lai killings break and the racism and vicious imperialism of America's frontier ideology would be laid bare for the world to witness (America 1998, 581–582, offers a transcript of Cronkite's report).

And the frontier was present elsewhere as well. Just weeks before Neil Armstrong and Buzz Aldrin landed on the lunar surface another version of that frontier was broadcasting its last original episode for the National Broadcasting Company. Television's *Star Trek* captured "Space, the Final Frontier," for what, in 1969, seemed like the last time. But in the years and decades to follow *Star Trek*'s popularity would continue to grow. Where the American military's involvement in Vietnam would seem to permanently scar America's frontier mythology, and where the initial popularity of the lunar missions would wane, *Star Trek* would continue to expand its popularity by remaking that frontier myth in an imagined and, at times, nearly utopian future.

Also in the fall of 1969, the American professor of physics Gerard K. O'Neill would begin teaching his freshman physics course at Princeton University by asking this question: "Is the surface of a planet the right place for an expanding technological civilization?" What would follow through his class in the fall of 1969 and into the following decade O'Neill would come to call the "High Frontier," and with it would come vision of orbiting space colonies fed with unimaginable amounts of free solar energy, expanded notions of personal freedom, and material plenty for all—both in orbit and back on the surface of the Earth. Like *Star Trek*, O'Neill's High Frontier was an imagined future of near-utopian ideals. Like NASA's Apollo missions, it was predicated on technological and scientific advances. And like America's involvement in Vietnam it was an attempt to push "American ideals" into yet another new location—into yet another new frontier.

1969: COMPETING IDEOLOGIES, COMPETING MYTHOLOGIES

The year 1969 was a year of competing visions of American frontier ideologies, then. Competing versions of frontier ideology and frontier mythology have been studied in multiple disciplines by many people. When those scholars claimed that they were describing the national universals of those myths or ideologies, they were, as often as not, merely describing differing aspects of a far more complex mythology than they envisioned. As the American diplomat and Yale University historian Robin W. Winks put it in this same period, "[t]he Frontier itself encapsulates the opposition of the good against the bad, the west against the east, the simple against the sophisticated, America against all others, and now nostalgia against change" (1971, 8). But strangely, it was also a year—and a period—in which examining such ideologies was not near the popular surface for either scholars or intellectuals. Historians, who had a long romance with the notion of the frontier as that environment that made Americans uniquely American, had been long attempting to give up on that paradigm and had certainly been unwilling to apply it to the 1960s. The originator of the "Frontier Thesis" among American historians, Frederick Jackson Turner, had been both held up and attacked by supporters and critics over the course of the seventy-five years since he had first proclaimed that "the frontier explained American development" in 1893. A year before the Apollo landings one of the most influential historians of the period, Richard Hofstadter, had spent a large section of his book *The Progressive Historians* attacking Turner for his errors, his vagueness, and his historiographical methodology in general. By the late 1960s, then, one of the grandest of the "grand narratives" of American historiography was no longer considered useful. Karen R. Jones and John Wills are correct in their reasoning: "Coined in 1893, the frontier thesis, somewhat predictably, appeared out of touch with the pressing social issues of the 1960s" (2009, 43). Indeed, any "grand narrative" of history seemed unworkable to a new generation of historians. As American historian M. J. Heale has noted in his discussion of why synthetic histories of the period are largely nonexistent, "the student can be introduced to most recognized topics in American history, such as the American Revolution or the New Deal, via essays that survey the available literature and lay out the scholarly battleground. Not so with the Sixties, for which few guides light the way" (2005, 133). Thus, while frontier rhetoric and frontier imagery abounded in the period, the disciplines that might have used it to collate a comprehensive history found such a project wholly objectionable.

For the discipline of professional history the period called "The Sixties" has continually resisted a historical synthesis of any kind according to Heale. The field of history fragmented in the period as much, if not more, than U.S. society itself. Heale describes this bluntly:

> One reason for this absence has been the difficulty of writing synthetic history as a consequence of the 1960s, when the powerful currents of

> race, gender, class, and culture undermined the older notion of a consensual society. *The pieces have never been put back together, again.* While these forces have fragmented the writing of American history in its entirety, for the Sixties an agreed narrative was never constructed in the first place. For an older topic like the New Deal, a reasonably coherent literature had come into existence within a generation. There were disagreements, of course, but the battle lines or the rules of engagement seemed to be fairly clearly understood. This could not be said of the Sixties, whose fate it was to be dissolved by the currents it spawned before a stable historiography could be written.
>
> (133; emphasis added)

The desire for a synthetic history has been a long-desired goal of professional American historians, and at least up to the 1950s, the Turner Thesis was a major option available to them. As the American intellectual historian John Higham put it, the Turner Thesis *was* American history up to the time of Turner's death in 1932, and then "for three decades from the 1930's to the 1950's, specialists in American history argued [the Thesis] fiercely" (Higham 1970, 118). By the 1960s, however, the desire for an all-encompassing synthetic history began to dwindle in American historiography.

If American historians were less than interested in a coherent narrative of the 1960s, much less one based on "the frontier," scholars in the field of American Studies were even less inclined to pay attention to it. Since the institutional emergence of American Studies just after the Second World War, the frontier had played a key role in shaping the perspectives of two generations of scholars. Where historians had worried about the empirical evidence of a real "frontier," American Studies scholars concentrated on the frontier as both symbol and myth. Taking the Turner Thesis—itself a product of the earlier generation of professional history in the Progressive Era—American Studies scholars adapted it significantly to fit the historiographical models of the time. Named by Higham as "consensus history," the use of the Turner Thesis was made to snugly fit the "strongly conservative movement to minimize the magnitude of social conflict in American history, and to depict a relatively homogeneous society" (1962, 23). But by the late 1960s, in the shadow of profound changes in both the American academy and American society, the unifying force that earlier had been attached to a consensus view of "the frontier" in American Studies gave way to the emergence of enquiries concerned with traditionally marginalized groups including African Americans, Hispanics, Native Americans, women, gays, lesbians, trans people, Catholics, and many others. Where the late 1940s through the early 1960s saw the American Studies version of the frontier useful in establishing broad outlines of a hoped-for genuine and singular American culture, a new generation of scholars were becoming more interested in those left out by such a paradigm, and profoundly uninterested, if not actually inclined toward a violent rejection, in the idea of a "singular American culture" at all. At "the very least," I have noted elsewhere, "the 'facts' used by a historian must be *perceived* by the reader of a work of history to be 'correct,'" and the idea of a singular frontier culture was no longer legitimate (Kapell and Elliott 2013, 358; original emphasis). It was not "true" to the generation of historians maturing in the 1960s and 1970s and, indeed, never had been. Furthermore, no longer were Americanist scholars interested in "consensus" at all, whether in history, American Studies, any of the traditional social sciences, or, indeed, any field of academic inquiry. Instead, the paradigms would come to be framed by what divided Americans, not what brought them together. Eventually termed *dissensus* by the Canadian Americanist Sacvan Bercovitch, any scheme that offered an analysis of the entirety of American society would be rejected out of hand by the early 1970s (1986). That meant, unquestionably, that the consensus version forwarded by the group in American Studies known as Myth and Symbol scholars would be anathema, and with it, most studies of "myth" itself would become verboten in academe.

Since 1969, these presentations of American frontier ideology have not gone unnoticed by later generations of scholars, however. The war in Vietnam has been the subject of countless works by political scientists, historians, and literary scholars. Similar groups have also examined NASA's Apollo missions with political historians leading the way. *Star Trek* is the subject of dozens of works by scholars of popular culture and cultural history. Political scientists and historians of science and technology examine Gerard K. O'Neill's High Frontier, although significantly less frequently than the other topics mentioned here. Prior to 1969, however, it would be easy to imagine an examination of each of these manifestations of frontier ideology as part of a broader examination of the continuing significance of the frontier in American culture. Since 1969 that has not been the case.

A simple explanation for this is one of definitional ease. It is possible to note President Kennedy's use of frontier rhetoric when declaring his new frontier and then turn to the far more obvious explanatory powers of Cold War ideology for an interpretation of both NASA's successes in the 1960s and its fading importance after the moon landing. An argument that centers on Kennedy's rhetoric need only confirm a long tradition of such images, and the NASA archives themselves will not mention frontiersmen or fur traders, cowboys, or continental expansion, but they will mention the Soviets. Turning to a direct engagement with American mythological structures of the frontier is to turn away from the empirical training historians are given. There will be no data in the archives, after all, about the culturally ephemeral world of "myth." And the period does not lend itself to such an analysis. The emergence of new fields and new paradigms, which were strongly surging by 1969, are not intended to examine broadly and shared cultural understandings. Even more established fields, such as history, were in the process of splintering into social history or political history or economic history or a wealth of other subfields, many of which managed to offer the word *new* into their approach. Emergent areas of study, such as minority studies programs, were organized specifically to examine those groups and individuals who were left out of previous "grand" narratives; they were not inclined to consider such narratives from any perspective other than how they had marginalized so many groups in the American nation.

As a result, social histories of the 1960s can be considered complete and offer extensive examinations of consequences of the Vietnam War domestically in the United States, yet in their "completeness" they can also totally ignore the Apollo moon missions. When literary and media historians point out the dialogue between, for example, *Star Trek* and the American popular understanding of Vietnam, little of NASA will remain. And while political scientists and now historians often examine the work of Gerard K. O'Neill those examinations tend to take place in a vacuum. Little of the influence of American culture, broadly or specifically, is left when one examines the imaginary space colonies dreamed up by O'Neill. But, in a way, that was O'Neill's intent, anyway: to leave the dissensus behind and to begin again.

"THE ZEITGEIST SCREAMED UNTIL IT WAS HOARSE": BETWEEN 1968 AND 1969

But, of course, the late 1960s as a whole can be considered a time of rapid change, and then similarly rapid retrenchment in America. It was, as the sixties activist, sociologist, and public intellectual Todd Gitlin put it, a

> [l]ong unravelling, a fresh start, a tragicomic *kulturkampf*, the overdue demolition of a fraudulent consensus, a failed upheaval, an unkept promise, a valiant effort at reforms camouflaged as revolution . . . whatever the image, the contending forces labored under a cloud of impending doom, or salvation, or both.
>
> (1987, 286–287)

This work offers a beginning in 1969, but if a single year were to be picked as the turning point of the late 1960s, most scholars would certainly not pick that year. Gitlin's *Kulturkampf*, undoubtedly, would be centered on 1968. Domestically, with the murders of both Martin Luther King and Robert Kennedy, with the height of the American numerical involvement in Southeast Asia and the tentative beginning of the long ending of the Vietnam War, multiple riots in major American urban centers, the declaration that the sitting President, Lyndon Johnson, would not seek reelection—the year is an easy place to assemble traditional political science or historical data as a moment of profound change. And, internationally and well as domestically, student protests surged across Western Europe as well as in the United States. Uprisings were seen in both Latin America and within the Eastern Bloc, controlled, primarily, by the Soviet Union. It is hardly surprising that historians Carole Fink, Philipp Gassert and Detlef Junker, for example, could edit a work called, simply, *1968: The World Transformed* (1998a). In fact, it is difficult to take umbrage with their project, which proclaims the "events of 1968 happened within national contexts yet took place across the globe—from Berkeley to Berlin, Bangkok to Buenos Aires, Cairo to Cape Town, Paris to Tokyo" (1998b, 1). Like perhaps no other single year in recent world history other that 1848 and, later, 1989 did a moment seem to indicate that profound changes were afoot—perhaps everywhere.

Among Americanists the "national context" of 1968 has also been recognized as a year of profound change. The Pulitzer Prize–winning historian Irwin Unger, writing with Debi Unger, could offer a similar reading domestically to that of *The World Transformed*. In their *Turning Point: 1968* they declare,

> Yet by any reasonable measure the changes crammed into 1968 were remarkable. It was the year when the Great Society petered out, when the civil rights movement stopped dead, when the liberal consensus of the sixties came apart, when the Vietnam War leveled off, when the cultural outsiders found their voice. Someone has described 1968 as the worst year of the century. As in everything else, it depends on where you stand. (1988, 532)

Irwin Unger saw such changes not just in the broad culture of the United States, but from his own position in academia as well. Just before the events of 1968 he had noticed changes in American History and among other Americanists, stating that the "notion of struggle as a central theme of our past," which had been ignored by historians seeking "consensus" for twenty years was, by 1967, reemerging. The new focus on conflict and struggle would be a major reason that Americanists would turn away from what they perceived to be the unifying influence of "myth" as an analytical tool. Unger saw the birth of "various New Left groups [that] have begun to feel the need for theory and analysis . . . [And has] already given birth to a New Left sociology, economics, and political science" as a key moment in the middle of the 1960s that would come to fruition in a 1968 that was "not just ferocious and tumultuous; it was also a year that changed America" (Unger 1967, 1238, 1237; Unger and Unger 1988, 3).

Thus, 1968 became, for Americanist scholars from many disciplines, journalists, and public intellectuals the year on which to focus. American historians Maurice Isserman and Michael Kazin are not contradicting any perceived wisdom when they declare "1968 was the pivot of the American decade" (2000, 221). American journalist and historian Jules Witcover could easily call 1968 the year that "both the vision and the dream" of an earlier 1960s "were truly shattered." It was, he wrote, best described as the "cataclysmic year of 1968" (1997, xii, xiii). Historian Douglas T. Miller, pondering the 1968 in the immediate aftermath of the assassination of Martin Luther King could state, simply, that "the prospect that a new politics of peace and racial justice might triumph and rebuild consensus in America faded" (1996, 230). Another historian, David Steigerwald, could offer that by 1968 the American involvement in Vietnam

> forced Americans to reconsider the most fundamental assumptions about the nature of their society: about its moral purpose, its strength or weakness of character, its historic role, its masculine attributes, the legitimacy of its political system.
>
> (1995, 118)

It should come as little shock that quotes such as the preceding come from books with titles or subtitles like *America Divided*, *The Year the Dream Died*, and the *End of Modern America*. Nor is it possible to suggest that journalists and historians such as the Ungers or Witcover are mistaken to offer 1968 as the year in which both change and the status quo seemed to collide in American culture.

But the other thing 1968 offers, in the words of the Amsterdam-born American journalist and novelist Hans Koning, is "a clarity of perception." Koning saw, retrospectively, 1968 as a year in which the "clichés, platitudes, and myths of our public life, what might politely be called our Fourth of July rhetoric, were suddenly seen as such, and not just by a handful of lefties in small magazines, but by the people at large" (1987, 13).

For Koning what had to come after 1968 was, in part, "rage and violence," but also "a consummation, an apotheosis even, of everything that had gone right, and everything that had gone wrong, in the year 1968" (1987, 181). Such an environment is culturally perfect for experiments with existing myths, which is precisely what began to happen in 1969. If 1968 was the moment, as Todd Gitlin would later put it, in which the "zeitgeist screamed until it was hoarse," then after that year was the time to examine what all that screaming was about (1987, 287). The year 1969 is a neglected year but, still, "in a single year, America saw the peaks and valleys of an entire decade—the death of the old and the birth of the new" (Kirkpatrick 2011, xiv). Examining that screaming, and reseeing 1969, mythologically, that is what this work is about.

DIALOGUES, DEBATES, AND DISCUSSIONS: TOPICS UNDER CONSIDERATION

This book, then, presents an analysis of the frontier ideology present in and around 1969 in the Vietnam War, the Apollo lunar missions, *Star Trek*, and Gerard K. O'Neill's High Frontier. As a synthesis of these four disparate visions of "the frontier," the examination here is historical and political, social and cultural. But, most important, it is an examination of these frontiers from a mythographical perspective.

Where previous analyses of the Vietnam War, for example, look to the conflict through the prism of Cold War geopolitics or the sociocultural ramifications of the conflict, here it is examined a one part of a continuing dialogue about the meaning, and the use, of frontier ideology and mythology. In a specific fashion, Vietnam seemed to undermine much—if not most—of the frontier mythology that had fueled it. The literary reactions to that war have certainly suggested as much. As a result Vietnam is offered here as an origin point for all the experiments in American mythology that would emerge in the period. Thus, where scholars studying the Apollo missions have focused on the domestic political motivations, the aspects of Cold War–inspired international relations, or the technological realities of the period, here it is seen as a different manifestation of frontier mythology, although one that is built, curiously, on images that long predate the frontier narratives presented in the late nineteenth century. A similar shift holds true for *Star Trek*, which has been most often offered as a simple extension of John Kennedy's New Frontier rhetoric. Here, instead, the willingness of the *Trek* narrative to specifically abandon notions of "paradise" that is frequently a part of frontier mythologies is highlighted. In a way, the narrative of *Star Trek*, for all its science fiction conventions, offered a far more realistic image of the "frontier" than NASA will ever manage. With the "closing" of the frontier by Frederick Jackson Turner in 1893 American culture has long searched for any possible "new" or "final" frontier, and it is one that *Trek* seemed to offer as the 1960s wound to a close. With the work of Gerard K. O'Neill another such frontier was offered as well. Although offering images similar to those of both NASA and *Star Trek*, however, O'Neill claimed his High Frontier could be available in the very near future. As a presentation of frontier mythologies that were essentially Utopian in nature, O'Neill's High Frontier offered Americans in the 1970s promises of a renewed culture.

Each of these topics has benefited from singular attention, but this work brings them together for a larger purpose. By examining them as a whole it is possible to see not a single version of American frontier mythology but an ongoing cultural debate about, to paraphrase the American pragmatic philosopher William James, the *varieties of the American mythographic experience*. Rather than seeing many manifestations of a singular "frontier myth," this work finds multiple versions of multiple myths. While those myths share much in common, they also diverge frequently and, on a rhetorical and cultural level, often forcefully. While they all center on ideologies of "the frontier," what those ideologies mean—and thus what the various myths of the frontier mean—they are also in constant dialogue. Previous scholars have noted that Vietnam and *Star Trek* entered into a protracted debate about the frontier in the late 1960s, but each also is in dialogue with NASA. NASA and the work of Gerard K. O'Neill also argue cogently in American culture about the possible meanings of "the frontier." Indeed, each topic in this study is in constant dialogue with the others, and the cultural debates inherent in this study are not new to American history or American Studies,

American social science or American literary history. This dialogue extends outward in both directions from 1969 quite easily, and through other topics, with continuing emphases and reworked ideas of what, mythologically, is meant by "the frontier" in American culture.

However, by focusing on these four topics, and extending out in both directions historically from 1969, it can be seen that this dialogue as key to understanding the American conceptions—*plural*—of itself as a nation and as a people. There is no intent, here, to suggest that these myths, together, comprise the whole of some national "American myth." Instead, it is an analysis of various myths, most explicitly active within the culture around 1969 that share common antecedents in American cultural and mythographic history. Where recent scholarship in American Studies chooses to emphasize plurality, diversity, and divergence and from that perspective correctly critique the attempts at "consensus" of earlier scholars, this study attempts to parse a careful line between these two extremes. Rather than offer "consensus" there is no attempt here to suggest other, equally important myths are nonexistent.

At the same time, however, there is an intentional purpose of not emphasizing the "dissensus" in contemporary scholarship either. If there is a unifying purpose here it is to suggest, though not declare in any sense, that while diverse voices have always existed within the various cultures of the United States, those voices have also frequently been engaged in an ongoing debate through mythical symbolism that is intended to lead to a conception, on some level, if not of a single people, then of a shared mythological vocabulary for dialogue. While that discussion—dialogue, debate, argument, ongoing conflict—has never come to a conclusion the purpose of a mythographic analysis is not to prophecy some ideal culmination but merely to assess what the purpose of the discussion, itself, is about. While the scholars after this period who have dealt with myth have largely done so by describing extensively negative connotations of mythic forms, this work further accepts American historian James Oliver Robertson's more nuanced approach that "our myths, whether they lead us to positive or negative responses—and they can do both—give us a sense that the world is understandable and explicable" (1980, 8).

It is a key point in this book that at best any attention to mythic constructs in American Studies today comes with a large dose of critical dissensus—that there now remains no kind of scholarship in American Studies that leaves open the possibility of there ever being a "whole" and, rather, scholars endeavor to Balkanize not only their field but also the nation-state that is the United States. It is the hope of this book that there can be a positive, and eventually inclusive kind of mythic fabric that makes the existence of the nation something other than an embarrassment to academics. But there is no intention that the topics discussed here, and the idea of a mythical frontier that connects them, are the only form of either frontier myths or, indeed, myths in general. This work is intended as an example of such forms—but is not meant to be exclusive in any sense. The consubstantial intent is to examine a set of frontier mythologies that were attempts at "new symbols of possibility" at precisely the moment in which American Studies and American historians were rejecting even the possibility of such analytic forms. But there remain, of course, many other symbols of possibility.

The four main topics considered here are both divergent and convergent, of course, yet all provide polysemous understandings of a mythology of the United States in the period that offered understandable and explicable visions of the world for Americans. Among the topics *Star Trek* stands out as a science-fictional version of American frontier ideology writ large, and there is a significant literature describing it as exactly that. However, the other topics have also been confronted through science-fiction narratives. As such, this work takes as part of its theme the mythographic notion presented by American myth scholar William G. Doty, namely, that "symbolic images and myths establish conservative benchmarks, but at the same time they anticipate *forms of the future* as they determine and shape ideals and goals for both individual and society" (Doty 2000, 26; original emphasis). *Star Trek* does this explicitly, and ultimately, this is how it is considered in this work. Yet *Star Trek* is not the only science fictional interpretation of the topics considered here. This is hardly surprising considering that American culture has often used the "imagined futures" of science fiction as a thought experiment for examining its own present. As Bronislaw Malinowski, the Anglo-Polish anthropologist who significantly influenced the American Studies school of Myth and Symbol scholarship, put it as long ago as the 1920s, "myth contains germs of the future epic, romance, and tragedy; and it has been used in them by the creative genius of peoples and by the conscious art of civilization" (1954, 143). This is especially true of the science fiction produced in the United States, and this study accepts American science fiction scholar and novelist Howard V. Hendrix's description that "the science fiction writer takes not the external world but rather, like the epic poet, his or her own personal vision of the future as the ultimate basis of his or her authority [and] in doing so, the science fiction writer engages in a form of visionary poetics" (1987, 43). It is not simply *Star Trek* that becomes a "form of visionary poetics" in a mythical sense here: writers have engaged, in one way or another, each of the topics this work considers. In considering the history, politics, and cultural influence of NASA, Vietnam, or the work of Gerard K. O'Neill, this work presents both traditional historiographical interpretations and such visionary poetics. But, in considering the science fiction influences within each of these topics this work attempts to recognize, in the words of science fiction scholars Robert Scholes and Eric S. Rabkin, that such work "concerns itself powerfully and continually with the examination of symbols central to our vision of our world and of ourselves" (1977, 169).

By allowing an examination of science fiction texts to work alongside the very real-world operations of NASA and the almost-plausible plans of Gerard K. O'Neill's High Frontier this work hopes, then, to find Doty's "forms of the future." In doing this the intent is to examine, as John Girling would have it, the relationship between myths, "which serve to legitimize the existing order," and "newly created myths, which are no less emotionally charged than their predecessors, may also inspire challenge to that order" (1993, 11). Joe Haldeman, considered a major writer and holder of the "Grand Master" award of the Science Fiction Writers of America has written science fiction novels dealing with both his experiences as a soldier in Vietnam and with Gerard K. O'Neill's High Frontier. Not coincidentally, as well, he also delivered *Star Trek* novels as part of his career, writing two in the 1970s.

If Doty's idea that " 'the mythic' codes for both the past *and* the future" is accepted, then an examination of how these reworked versions of frontier myths are imagined by some writers as part of that future can help explain the very present from which they were writing (Doty 2000, 92; original emphasis). This work accepts what Doty describes as a needed examination of those myths that too often have been "left hidden [as] the many political assumptions that have led the United States to its outrageous Manifest-Destinying across the hemispheres, a policy that has led to worldwide chaos in many of our lives and those of the nations we have felt obliged to make more like us" (1995, 211). Recognizing, as many contemporary scholars have, that this is indeed a problem with the "political assumptions" inherent in such an unexamined mythology, this work also takes the topics here as an example of the general culture actually examining its own mythology. So, this work derives its cue from the perspective of British religious historian Karen Armstrong, who noted "that every time men and women took a major step forward, they reviewed their mythology and made it speak to the new conditions" (2005, 11).

Thus, by bringing together the Vietnam War, NASA's Apollo missions, *Star Trek*, and Gerard K. O'Neill into a synthesis of mythographic analysis the intent is to provide two things: a new examination of this specific period in American cultural history and a new model for examining that cultural history more generally, in other moments, and other times.

Such a study, of course, must begin with a specific and careful presentation of how contemporary scholars employ the term *myth*. The reasons for offering such a presentation, however, are complex. By the middle of the 1960s both historians and American Studies scholars had begun to fully shun models of "the frontier" as having utility in understanding American culture or history and in part this was due to the dominant ways in which both fields presented "myth" as an analytical tool suitable for examinations of cultural or historical realities. This work considers a number of scholarly movements that have declared themselves "new" in one way or another, including "New Western Historians," "New Americanists," "New Left historians," and others. Strangely, these movements have been content to consider changes in their fields "new" but have rarely bothered to examine the study of myth to see if similar "new" variants have emerged. As a result recent Americanist scholarship has tended to avoid using myth or, when mentioning it at all, has tended to ignore the development of myth scholarship in the last forty years. Thus, the "New Americanist" Philip Fisher can describe myth as "fixed" and "stable," two terms that would not be acceptable descriptions in contemporary myth scholarship (1991, vii). And the British American Studies scholar Neil Campbell, drawing from French poststructuralism to explain an American "global West" could offer that myth has a "gridlike neatness" of a "rooted, treelike . . . structure" (2008, 7). Again, contemporary myth scholars would reject such a hierarchical conception of their topic of study. To fully engage with the many frontiers of 1969, then, it is necessary to use a more pragmatic and contemporary view of "myth" and mythographic analysis. Contemporary scholars have rejected "myth" because of its nonempirical aspects, or in a desire to examine what they consider the more transformative potential of rhetorics, though both critiques are built on a fundamental misreading of the purposes of myth, itself. Thus, this project presents a model of "myth" built on poststructuralist and postmodern theories, but firmly attempts to avoid the debilitating jargon of such works. To do this, and still result in a working paradigm of "myth" that is useful for specifically American cultural analysis, the model of myth presented here is built on pragmatic conceptions of "truth" as applied to mythology. Finding "mythological truths" in the frontier myths of 1969 is not to discount the more traditionally empirical methods of historians and political scientists or the rhetoric-based approach of contemporary American Studies scholars. But it is possible to move beyond those analyses and attempt to clarify a more complete set of images of the American cultural experience in the period when bringing to bear a more pragmatic version of precisely what can be meant by "myth."

With such a model of "myth" offered an examination of the initial conception of Frederick Jackson Turner's Frontier Thesis and of how historians used that thesis as an explanatory model, can be presented. Doing so after a discussion of myth is quite intentional, for Turner can be best read as a mythographer using the terminology of a social scientist. Tracing the Turner Thesis from emergent concept, to dominant paradigm, to dismissed theory is an examination of movements toward the moment when the frontier ideology of 1969 could go unexamined. As the Turner Thesis began its steady decline as a paradigm of empirical history during the Great Depression and the Second World War, American Studies emerged to use Turner's ideas not for descriptive, empirical history but as cultural history. Thus, it is necessary to show how scholars of American Studies built on the Turner Thesis with notions of myth and symbol to provide a somewhat different explanatory model of American culture. Working alongside the postwar "consensus history" the Turner Thesis nicely dovetailed into a model for understanding why Americans felt unique and why they felt *uniquely American*.

With such a mythographic model of analysis in place, combined with a rough sketch of earlier forms of myth studies in various fields related to this work, it is possible to turn to Vietnam, NASA, *Star Trek*, and O'Neill and see how each of these cultural events carried with them notions of "the frontier" that were both fully recognizable by Americans and fully engaged with the broader realm of American myth. At the same time, the differing versions of "frontier myths" inherent in NASA, *Star Trek*, Vietnam, and O'Neill's High Frontier complement and contradict each other. In other words, it is possible to parse the mythological differences in various versions of "the frontier" as those versions remain in constant dialogue with the others and, often, with themselves. In 1964 American Studies scholar Leo Marx, near the end of his *The Machine in the Garden*, offered that for American culture to move beyond such myths of the frontier would "require new symbols of possibility." Not coincidentally for the present work, speaking at a conference at Wayne State University in Detroit in 1969 Marx attempted to defend his method of analysis, and the method of his American Studies colleagues, to a group of traditional historians. He argued that there was no aspect of his approach "which can be called objective" but that it was still a search for a cultural significance which had a "compelling truth value" (1969, 89; Marx's talk was originally presented in May 1969, at a conference on foreign policy and public opinion organized by Mel Small of the university's history department). For a time the Vietnam War seemed to destroy any "compelling truth value" of American frontier myths. But, rather than search for "new symbols of possibility," eventually the traditional myths were modified in various ways. The experiments with those myths are the purpose of this work as aspects of American culture began to offer differing ways of utilizing traditional American symbols in an attempt to find Doty's "forms of the future." As a result the continuing question asked by American culture—or various American *cultures* if one prefers the plural—becomes central to the ongoing story of an American nation, whole or splintered, complete or conflicted and it remains Caroline Kirkland's question about the limits of the idea of "the west" or "the frontier." The question becomes: Just where, exactly, is the *Next Frontier*? And will it, as many Americans still firmly believe of their mythical past frontier, be *Exceptional*?

NOTE

1 References in this introduction consist only of specific quotes if sources are quoted in main chapters of the text. Further information, and citations, will be provided in the chapters that follow.

WORKS CITED

America, Library of. 1998. *Reporting Vietnam: American Journalism 1969–1975*. Des Moines: Library of America.

Armstrong, Karen. 2005. *A Short History of Myth*. Edinburgh: Canongate.

Bercovitch, Sacvan. 1986. "America as Canon and Context: Literary History in a Time of Dissensus." *American Literature* 58: 99–107.

Campbell, Neil. 2008. *The Rhizomatic West: Representing the American West in a Transnational, Global, Media Age*. Postwestern Horizons. Lincoln: University of Nebraska Press.

Cronkite, Walter. 1969. "Apollo 11 Special Coverage." *Moon Landing*. New York: CBS.

Doty, William G. 1995. "Silent Myths Singing in the Blood: The Sites of Production and Consumption of Myths in a 'Mythless' Society." In *Picturing Cultural Values in Postmodern America*, edited by William G. Doty, 187–220. Tuscaloosa: University of Alabama Press.

———. 2000. *Mythography: The Study of Myths and Rituals*. 2nd ed. Tuscaloosa: University of Alabama Press.

Fink, Carole, Philipp Gassert, and Detlef Junker, eds. 1998a. *1968: The World Transformed*. Publications of the German Historical Institute. Cambridge: Cambridge University Press.

———. 1998b. "Introduction." In *1968: The World Transformed*, edited by Carole Fink, Philipp Gassert, and Detlef Junker, 1–27. Cambridge: Cambridge University Press.

Fisher, Philip. 1991. "Introduction: The New American Studies." In *The New American Studies: Essays from Representations*, edited by Philip fisher, vii–xxii. Berkeley: University of California Press.

Girling, J. L. S. 1993. *Myths and Politics in Western Societies: Evaluating the Crisis of Modernity in the United States, Germany, and Great Britain*. New Brunswick, NJ: Transaction Publishers.

Gitlin, Todd. 1987. *The Sixties: Years of Hope, Days of Rage*. New York: Bantam Books.

Heale, M. J. 2005. "The Sixties as History: A Review of the Political Historiography." *Reviews in American History* 33 (1): 133–52.

Hendrix, Howard V. 1987. "The Thing of Shapes to Come: Science Fiction as Anatomy of the Future." In *Storm Warnings: Science Fiction Confronts the Future*, edited by George Edgar Slusser, Colin Greenland, and Eric S. Rabkin, 43–54. Carbondale: Southern Illinois University Press.

Higham, John. 1962. "The Construction of American History." In *The Reconstruction of American History*, edited by John Higham, 9–24. London: Hutchinson & Co.

———. 1970. "The Divided Legacy of Frederick Jackson Turner." In *Writing American History: Essays on Modern Scholarship*, 118–29. Bloomington: Indiana University Press.

Isserman, Maurice, and Michael Kazin. 2000. *America Divided: The Civil War of the 1960s*. New York: Oxford University Press.

Jones, Karen R., and John Wills. 2009. *American West: Competing Visions*. Edinburgh: Edinburgh University Press.

Kapell, Matthew Wilhelm, and Andrew B. R. Elliott. 2013. "Conclusion(s): Playing at True Myths, Engaging with Authentic Histories." In *Playing with the Past: Digital Games and the Simulation of History*, edited by Matthew Wilhelm Kapell and Andrew B. R. Elliott, 357–69. New York: Bloomsbury Academic.

Kirkland, Caroline. 1855. *A New Home—Who'll Follow? Or, Glimpses of Western Life*. New York: C.S. Francis and Co.

Kirkpatrick, Rob. 2011. *1969: The Year Everything Changed*. New York, NY: Skyhorse Publishing.

Koning, Hans. 1987. *Nineteen Sixty-Eight: A Personal Report.* New York: Norton

Malinowski, Bronislaw. 1954. "Myth in Primitive Psychology." In *Magic, Science and Religion and Other Essays*, 93–148. New York: Doubleday.

Marx, Leo. 1964. *The Machine in the Garden: Technology and the Pastoral Ideal in America.* New York: Oxford University Press.

———. 1969. "American Studies. A Defence of an Unscientific Method." *New Literary History* 1 (1): 75–90.

Miller, Douglas T. 1996. *On Our Own: Americans in the Sixties.* Lexington: D.C. Heath.

O'Toole, Thomas. 1969. "The Eagle Has Landed." *Washington Post*, July 21, sec. A.

Prochnau, William W. 1969. "Man Walks On Moon." *Seattle Times*, July 21, sec. A.

Robertson, James Oliver. 1980. *American Myth, American Reality.* New York: Hill & Wang.

Scholes, Robert, and Eric S. Rabkin. 1977. *Science Fiction: History, Science, Vision.* New York: Oxford University Press.

Steigerwald, David. 1995. *The Sixties and the End of Modern America.* New York: St. Martin's Press.

Unger, Irwin. 1967. "The 'New Left' and American History: Some Recent Trends in United States Historiography." *The American Historical Review* 72 (4): 1237–63.

Unger, Irwin, and Debi Unger. 1988. *Turning Point: 1968.* New York: Scribner.

Wetzels, Walter D., ed. 1973. *Myth and Reason: A Symposium.* Austin: Published for the Dept. of Germanic Languages of the University of Texas at Austin by the University of Texas Press.

Wilford, John Noble. 1969. "Astronauts Land on Plain; Collect Rocks, Plant Flag." *New York Times,* July 21, sec. A.

Winks, Robin W. 1971. *The Myth of the American Frontier: Its Relevance to America, Canada and Australia (Sir George Watson Lecture, Delivered in the University of Leicester, 21 January 1971).* Leicester, UK: Leicester University Press.

Witcover, Jules. 1997. *The Year the Dream Died: Revisiting 1968 in America.* New York: Warner Books.

2 Paradigms Lost and Paradigms Regained

A Mythography of the Lost Frontier

> [The] implicit claim to be antimythical and purely scientific is itself a myth—that is, a set of unconsciously held, unexamined premises.
>
> —Robert Jewett and John Shelton Lawrence, 1977 (Jewett and Lawrence 1977, 17)

> Myth and history are close kin inasmuch as both explain how things got to be the way they are by telling some sort of story.
>
> —William H. McNeill, 1985 (McNeill 1986a, 3)

The terms *myth* and *frontier* occur so often in this work that readers deserve some warning about their use.[1] The words were initially used in a generally positive way to explain a kind of "national mind" of the American people and, later, were used critically to highlight various levels of racism, sexism, and imperialism, found in American culture. In the case of both words I intend them as terms used with awareness of that genealogy. Where each term has come to represent profoundly negative concepts fused with similarly negative kinds of historical facts, scholars have repeatedly rejected them or have used them only as metonyms for uniformly negative aspects of the American past. However, both were jettisoned from areas of academic thought when various parts of U.S. culture were engaged in reimagining them. The combined term of *frontier myths* followed a similar pattern of noncritical use through the 1960s, then expressly critical use from the 1970s onward. Eventually the idea of frontier myths was supplanted by the term *rhetoric*. This new scholarship was confined primarily to English departments despite the fact that earlier generations of Americanists employing these concepts were avowedly interdisciplinary.

In the areas of American Studies and history, using frontier myths in scholarship have had a decided influence on the practical affairs of scholars' careers. It is partially because of this that the intellectual constructions of such myths were becoming less popular at the exact moment they could have been positively employed in scholarship. But also at issue was the simple fact that the idea of such myths was but one of many positions taken in which a single aspect of a "distinctive American national culture" explained a proverbial kind of American uniqueness or Exceptionalism. Richard P. Horwitz (2001) described the academic situation nicely, distilling it to this:

> In documenting that distinctiveness [of America] they cite a bewildering array of influences: God's grace, Puritan theocracy or the separation of church and state, early colonists' military might or their resistance to diseases that they spread, the timing and composition of particular waves of immigration, the "availability" of arable land, continental abundance or regional shortages of resources, free enterprise or slavery, the spirit of science or unfettered individualism, political liberty or the suppression of dissent, technological prowess or omnivorous consumerism, mobility, individualism or conformity, pragmatism or idealism, relative peace and prosperity or racism and violence. Whatever the explanation, at issue has been less whether there *is* anything distinctly "American"—a way of life uniquely associated with a population, a setting, and a nation-state—than the best way to describe and evaluate it.

Frontier myths were one of a possible array of singular and unitary explanations for the idea of American uniqueness and were rejected when many such models were rejected or found to be consequentially negative.

This work argues that there never has been a single "frontier myth" but actually many of them.

Additionally, at the precise moment when "frontier myths" were being found increasingly unacceptable to a generation of scholars and used with increased negativity Americans were in the process of experimenting with new forms of frontier myths that went largely unexamined because of the institutional momentum against the very idea of such myths. This was especially true if those myths had a potentially positive meaning. Because scholarship in the 1970s and after refused to see anything but the critical and the negative in notions of frontier myths they missed the "deeply held conviction" that lead to positive frontier myths on television, in movie theaters, at Cape Canaveral, and in utopian planning for a future in space.

To understand how this work hopes to academically resuscitate frontier myths in different forms than they have previously been used it is necessary to understand, somewhat, why the terms were rejected in the first place. In the areas of foci touched upon here, then, it is imperative to consider the rejection of "myth"—and, specifically, the "Myth and Symbol school" in American Studies—since that is where the rejection was most forceful. The next requirement is to understand the rejection of the concept of a "frontier" in American history for it is there that the rejection, too, was most vehement.

For American Studies, the term *myth* came to be rejected by scholars seeking to find a more empirical center for the field as well as scholars trained to find the term indicative of too much that was negative in the American past. To reject myth they considered the term as it was employed by non-specialists within their field and found the term wanting. Or, when used, they found it to ignore far too much that was sexist, racist, and imperialist. In short, they constructed a straw man out of the previous—and poorly defined—use of the term by American Studies scholars and then had little difficulty rejecting not just that version of "myth" but, rather, all potentially positive myths of any kind.

At almost the same time from the 1960s through the 1990s American historians were right to reject the term *frontier* because if its origin and development. This genetic fallacy allowed them to note the inherent imperialism, racism, sexism, and a host of other –isms found in the original conception of the Frontier Thesis and reject the notion of using the idea of "frontiers" in any way to understand the United States. Historians' rejection of a frontier was a rejection of its nonhistorical underpinnings, but the result became that the idea of a "frontier" in contemporary cultural understandings was, itself, rejected as well.

In both cases the concepts of myths and frontiers were rejected by scholars at exactly that moment when myths of the frontier were once again morphing into new forms in American cultural discourses. Since previous scholars in American Studies used a simplified idea of myth somehow all mythic analyses were rejected. Since those scholars saw myths as offering conservative cultural cohesion, if they did look to myths they saw them as jaundiced, withered and primarily having negative effects in light of new needs in U.S. cultures. And, since previous generations of historians had used *frontier* in ways that implicated it in various rejected –isms, the very notion of a frontier hypothesis had to be jettisoned as flawed even if most Americans had their own versions of a past that included the idea in ways that affected their daily lives.

This book argues that the rejection of *myth* was based on an antiquated definition of that term and that the rejection of *frontier* was based on a reasonable reaction to the racist, sexist, and imperialist underpinnings of its use, even though everyday Americans were often fully invested in the idea of frontiers in the American past. However, because the two terms were rejected for differing reasons, the idea of frontier myths became bound up in such gloomy arguments as well. To employ these terms, then, we have to present a more coherent, complex and malleable form of *myth* while considering a more mythic conception of the American cultural idea of "frontier," or of "the West." The idea of the frontier—like myth itself—was never static. It was a constantly shifting idea, a continually restricting semiotic web defined for individuals, for groups, and for the American nation as both pragmatically useful and mythically transformative. We must try, then, to use the terms not as academics have previously employed them, but (to steal a phrase) as "new symbols of possibility" (Marx 1964, 365).

I begin with "myth" and then move to the idea of a "frontier," not as a category of American history but as one of American myth.

A REKINDLING OF MYTHOGRAPHY FOR AMERICANISTS

The Myth and Symbol school, which dominated American Studies from its origins until the late 1960s operated with a specific—and rarely defined—idea of what they meant by myth. What was wrong about their myth conceptions was their initial formation, which was rarely examined later, of the term. In the broadest sense, they would agree with the following statement by the Anglo-Polish anthropologist Bronislaw Malinowski on myth, even though he is speaking of "primitive" culture:

> Myth fulfils in primitive culture an indispensable function: it expresses, enhances, and codifies belief; it safeguards and enforces morality; it vouches for the efficiency of ritual and contains practical rules for the guidance of man. Myth is thus a vital ingredient of human civilization: it is not an idle tale, but a hard-worked active force; it is not an intellectual explanation or an artistic imagery, but a pragmatic charter of primitive faith and moral wisdom.
>
> (1954, 101)

Malinowski describes the use of "myth" in an attempt to ascertain how myth functions *within* a society for the promotion of the stability of *that* society. Other than this indication of the primitive, however, Malinowski's purpose was to uncover the function of myths—that is, what they *do*. Examining myths, Malinowski believed, would enable an understanding of the relationship between "the word, the mythos," of a culture and that culture's "moral deeds, their social organization, and even their practical activities" (Malinowski 1954, 96). Malinowski described myths, as mythographer William G. Doty has put it, "as turning into a societal glue, undergirding . . . social self-understanding and hence worldview" (Doty 2000, 67–68). In the functionalist approach to myth, again in Doty's words, "Malinowski understood myth to be a direct statement of social realities" (Doty 2000, 131). His intent was to examine myths as a part of the "social function of maintaining and *expressing solidarity*" (Cohen 1969, 343; emphasis added) Eric Csapo put this even more bluntly, stating that for Malinowski, myth "justifies the social order, its institutions, practices, customs, and moral codes" (Csapo 2005, 142). Malinowski's definition of myth was thus essentially *conservative*. It worked not to provide options for individuals within a society, but to limit those options and, by limiting them, to limit change as well. Eventually, this conservative—and dated—understanding of myth will be the one that contemporary scholars in American Studies have in mind when they reject their field's earlier attempts at examining an American mythos.

Percy Cohen was fully correct in his assessment of Malinowskian functionalist interpretations of myth. The paradigm would hold sway over much of social scientific study in the early to mid-twentieth century, meaning that for most scholars "myth . . . has to be defined by its function" (Malinowski 1963a, 249).[2] That means the study of myth is the study of "a body of narratives woven into [a] culture, dictating belief, defining . . . ritual, acting as the charter of . . . social order and the pattern of . . . moral behaviour" (Malinowski 1963a, 249). For both individuals and societies, then,

> the main cultural function of mythology is the establishment of precedent . . . the binding forces of morals and law, of the real value of religious ritual, by a reference to events which have happened in a dim past . . . Myth, coming from the true past, is the precedent which holds a promise of a better future if only the evils of the present can be overcome.
>
> (Malinowski 1963b, 291–292)[3]

The functionalist interpretation of myths that Malinowski attempted to describe were parts of a larger lived "cultural context," what he called "the ideas, emotions, and desires associated with a given story," that, together with "a retrospective pattern of moral values [and] sociological order" would show that "myth is, therefore, an indispensable ingredient of all culture" (Malinowski 1954, 145–146). Midcentury myth theorists provided a definition of their term that, as Doty noted ironically, was "a sort of . . . SuperGlue that could counter the social fragmentation imposed" by modern nation-states (Doty 2001, 3).

To put it another way, the study of myth as "ideas, emotions, and desires" offers the study of "larger or smaller units of the same kind of thing, namely an intellectual construction that fuses concept and emotion into an image . . . [and] sometimes exert a decided influence on practical affairs" (Smith 1971, xi). That famous opening line from the first edition of Henry Nash Smith's myth and symbol classic *Virgin Land* composes much of both the American Studies version of myth and that of the early-twentieth-century functionalists, characterized by Malinowski. Smith's study of dime novels, Leatherstocking, Buffalo Bill, church sermons, and newspaper editorials very much operates from the perspective, in Malinowski's phrase, that "these stories form an integral part of culture," (Malinowski 1954, 108). For Smith the myths he examined functioned to hold American society together, and they were, thus, essential and conservative in their function.

What scholars at the time would have called the "civilized" culture of the United States, and what today would be characterized as "complex" could, Myth and Symbol scholars asserted, be known "holistically" through their dominant myths. So, where anthropologists attempt to include myths, symbols, and rituals as part of a broad "holistic" undertaking of a less complex culture made up of many different parts, American Studies scholars reversed the method almost completely. Now the "nonsimple" or complex culture of the United States could be understood wholly—that is, holistically in American Studies terms (though not anthropologically)—through its myths. As such, the American Studies scholar Cecil F. Tate is fully correct in saying that

> [a]s American Studies understands myths, they are inextricably involved with a particular national culture, a whole culture, from which they emerge and whose "dominant thought forms" they bring forth with imaginative impact. Whether or not we agree with this vision of culture, it has had a decisive impact on the development of American Studies.
>
> (1973, 24)

In other words, Myth and Symbol scholars offered a narrow definition of myth, which implied an equally narrow definition of culture. But this is not surprising since those scholars were neither trained in religion (and thus myth) nor anthropology (and culture).

The American scholar of religion Ivan Strenski has described this phenomenon among scholars who use the term *myth*. There are, according to Strenski, those who "deal with our basic assumptions about myth: they chiefly *prescribe* what the concept of 'myth' is and how it should therefore be studied and used" (1987, 2; original emphasis). Strenski calls these the "first class of theorists" about myth. His "second class" is a more difficult and diverse group. They are "those applying the term 'myth'" and they

> seek primarily to *use* ideas of myth: they are intentionally innocent of theory and satisfied to employ 'myth' as it is available from our cultural and theoretical stocks without necessarily having to probe conceptual foundations.
>
> (Strenski 1987, 2; original emphasis)

Of these two "classes" of myth scholars those in American Studies—or any related field but using the term *myth*—are most often in the latter. However, the problem here is not quite that simple. Scholars in the first class, as Strenski notes, rarely enter into a sincere dialogue with others of that same group. "They have little in common but the word 'myth,'" Strenski argues (1987, 3). For scholars working in the American Studies Myth and Symbol school it is not surprising that they do not carefully theorize their use of the term *myth*, as actual scholars of myth, itself, rarely accomplish this feat either. The "theorists of myth," Strenski notes, "so systematically avoid reference to each other . . . [that] in the rare cases where conversation does occur, it follows the rule of talk at cross purposes" (1987, 3). Social anthropologist Percy S. Cohen offered a state of the field of myth studies in, importantly for this book, 1969. He offered a more generous explanation of the phenomenon: "There are many theories of myth, but they are not rival theories," he claimed. "Different theories often explain different statements about myth" (Cohen 1969, 338). Yet Cohen noted much the same thing as Strenski, that "there is an inevitable confusion in the debate between different proponents of different theories" of myth (Cohen 1969, 338). In other words, as Laurence Coupe has put it, "myth is paradigmatic, but there is no true paradigm" (2009, 5). Thus, American Studies scholars fall into a similar trap as do all "second class" myth scholars: they remain likely to employ a term in divergent fashions and resultantly "talk at cross purposes" about the very object of their studies—namely, "myth." Because of this problem it has been fairly simple for later Americanists to dismiss the work of previous Myth and Symbol scholars by simply offering a more precise, or simply different, definition of "myth" and then refuting earlier definitions as being unable to cope with the needs of academic analyses.

It will become apparent in the remainder of this chapter that this "talking at cross purposes" between mid-century definitions of *myth* and how those definitions found their way into American Studies remains *the* fundamental flaw in earlier Myth and Symbol scholarship—though calling it a flaw is somewhat unfair, nonetheless. Myth and Symbol scholars used a definition that originated in functionalist anthropology and used it well. That their eventual critics were unable to find, ironically, a function in that work is to merely note that scholarly work evolves over time.

However, we can begin to understand the idea of myth in Myth and Symbol scholarship by contemplating the mythography of the original Frontier Thesis.

MISTAKING MYTH FOR HISTORY: FREDERICK JACKSON TURNER'S CONTINUING SIGNIFICANCE

In 1893 Frederick Jackson Turner, then a young professor of history from the University of Wisconsin, provided a way to envision the American past around the term *the frontier* as a holistically complete system. It was different from the European nations from which he believed it sprang and had evolved along its own unique path, and Turner claimed that because of *this fact*, it was significant. Standing before a gathering of professional historians at the American Historical Association's annual meeting in Chicago, Turner presented a thesis of American history that would reverberate for more than a century since.

That 1893 paradigm, stated in "The Significance of the Frontier in American History," offered a new model for explaining the unique forms of American democracy and culture, yet he utilized the acceptable social scientific rhetoric of the era (Turner 1999). The "Frontier Thesis" or the "Turner Thesis," presented a model that seemed to explain both how and, more important, *why* America differed from other nations. He put it bluntly: "The existence of an area of free land, its continuous recession, and the advance of American settlement westward, explain American development" (Turner 1999, 19). However, as one of the few scholars to approach the topic of frontier mythology after the 1960s, Richard Slotkin, has noted, the "association of westward expansion with an ideology of American uniqueness and world mission" is not original with Turner or other historians of his time but easily "traced back to Puritan concepts" as well as notions of Manifest Destiny (1981, 610). Indeed, the Turner Thesis is, on a certain level, merely a historical restatement of what mythographer Richard T. Hughes calls "the primal meaning" of America (2003, 2).[4] Thus, the genealogy going back to the Puritans and traced through nineteenth-century concepts of Manifest Destiny is useful, the historical interpretation of America the Turner Thesis offers makes it perhaps the single most important presentation in all of American historiography. The debates surrounding the Turner Thesis have raged, literally, for much of the period since his initial presentation, and those debates are only important here because of how the reworked thesis eventually became part of the "Myth and Symbol" school of American Studies. Nonetheless, it remains a necessary step in understanding how the Turner Thesis captured the zeitgeist of Turner's time.

Furthermore, while scholars in the 1960s happily turned away from the thesis, Turner's ideas have nonetheless remained mythologically central to understanding the many cultures of the United States. The essential point in a re-presentation of the Turner Thesis, then, is not to find its historical errors, but to note that the critiques directed against it are the criticisms of empirical historians. But, from the perspective of myth, those very critiques actually point toward the most significant aspect of the thesis. When the question of empirical accuracy is left aside, in other words, historians' empirical scrutiny actually reveals Turner's mythic importance.

What Turner suggested, simply, was that the "free land" of an empty North American continent allowed settlers to regress from their overly civilized European roots, reencounter the primitive, and from that process American democracy was born. Building his argument from census data, demographics, and economics, Turner was attempting a major shift in historical thought for the time. He was both changing the perspective of how U.S. history was viewed and of how history, itself, was undertaken. As William Cronon aptly described it, Turner was attempting to "transform history from literature into science" (1991, 74). At the forefront of a new, professional generation of historians, Turner was setting forth to make history a scientific and empirical enterprise.

Put bluntly, Turner was "applying the techniques of social science . . . to the settlement and social evolution of the United States, particularly to the region beyond the Appalachians" (Cronon 1991, 81). Another way of conceptualizing Turner's work is that he was attempting to answer Caroline Kirkland's question about the limits of "the West." He did, after all, offer a detailed idea of those limits—or, in his own term, of "the frontier"—and, in doing so, stated that the frontier he was describing, according to his data, was no longer in existence. As an historian Turner offered an interpretation of the American past. But as an unintentional mythographer he also offered that explanation as the key to the present. As David W. Noble described it in the middle of the 1960s, the thesis may have offered an examination of the American past to his audience in Chicago, but in reality he "explained the cultural dilemma of the country in the 1890s" (Noble 1965, 37). For Noble, in the title of his book, this makes Turner one of the *Historians against History* and is a central flaw in the thesis. But, describing the present with reference to the past is at least one *purpose of myth*. Thus, mythographically and mythologically it is not a flaw but a strength.

At the time the Thesis fundamentally moved American history away from Europe and toward an entirely new focus. As Richard W. Etulain described the thesis, Turner provided a "provocative call to understand American identity as primarily the result of shifting national frontiers, advancing from the Atlantic to the Pacific Coast" (1999, 4). Turner did this by emphasizing those experiences of the West that were part of his Wisconsin childhood. Turning fully away from European precursors for American development, Turner looked to the experiences of farmers and traders. It was, as Etulain notes, a thesis built on autobiographical detail:

> Because of [Turner's] own background and his attachment to agricultural communities, farmers were the central figures in his frontier story. True, explorers, fur trappers, railroad builders, stock-raisers, and townsmen were part of the westward movement traced from East to West, but the hard-working farmers were the major protagonists of his histories.
>
> (1999, 5–6)

Turner built a singular thesis for explaining all of American development not on the elite of the East Coast but on the common people who moved continuously westward. Thus, it was Turner who first presented a specific thesis for organizing America's frontier experience in an understandable form. It was Turner who promoted that thesis for decades. And it is Turner who remains, more than one hundred years after the first declaration of his thesis concerning the American frontier, a controversial enough figure among American historians to still be debated. The details of the Turner Thesis have entered American culture directly and often indirectly as no other academic presentation before or since.

Thus, a discussion of the Turner Thesis is, almost by necessity, as much a discussion of the American mythos as it is a discussion of American historiography.

Among all contemporary American historians it is perhaps William Cronon who most carefully summarizes these many effects of the Turner Thesis. For Cronon what Turner did was use the idea of "the frontier" to explain a "place where easterners and Europeans experienced a return to a time before civilization when the energies of the race were young" (Cronon 1987, 157). Turner's frontier, in the accepted social scientific parlance of the day, became a shorthand method of understanding American Exceptionalism. Although Turner did not originate an idea of American uniqueness, he did present a new synthetic model for understanding it. Turner claimed that through the frontier, as Cronon explains it, allows a "descent to the primitive," and once that was complete the American people would find

> frontier communities [which] underwent an evolution which recapitulated the development of civilization itself, tracing the path from hunter to trader to farmer to town. In that process of descent and revolution—as the frontier successively emerged and vanished—a special American character was forged, marked by fierce individualism, pragmatism, and egalitarianism. Thus, fundamentally transformed as a people, Americans built their commitment to democracy, escaped the perils of class conflict, and overran a continent.
>
> (1987, 157)

The Turner Thesis allowed for an explanation of American Exceptionalism because "the frontier experience had facilitated the growth and development of democracy and had been the source and sustainer of almost every beneficent facet of the American character" (Wrobel 1993, 37). Turner's claim for America uniqueness was based, in a simple and logical way, on the uniqueness of the American frontier. Even this was not a particularly new concept, however. David W. Noble noted that such a distinction had been central to both American scholarship and American literature since the early nineteenth century. "The conflict between the idea of nature and the idea of civilization," Noble claimed, "is central to the philosophical concerns . . . of the United States as either a state of nature or a continuation of European civilization" (1968, x). In describing such a dialectic, however, Turner used non-specific and idealized concepts, including undefined terms like *social forces*, or *composite nationality*, and even *the frontier* to create a vision of the American past that set it out as unique in every way. And, while "Turner's argument was insufficiently analytical, and his key frontier concept so ill defined that it could be applied to virtually any historical circumstance," he nonetheless *did* create a unique vision of American history (Cronon 1991, 90). In the sense that Doty means "mythic accounts, especially those that narrate beginnings, embody ideas of wholeness, or order replacing chaos," the thesis offers a comprehensive mythic understanding of both the American past and present (Doty 2000, 60). To enter that vision of the Turner Thesis, as Patricia Nelson Limerick has put it, was to

> enter an enchanted world . . . [of] abstractions [that] are tangible and virtually animate, right on the verge of speaking for themselves. Conditions, forces, ideals, institutions, traits, types, elements, and processes

> inhabit the Turnerian world like the weightiest and most settled of citizens. If you bump into a social force or a pioneer ideal, it will be you who gets the bruise.
>
> (1995, 697)

Turner hoped his thesis would be a straight and true objective history, but it was bound up in too many abstractions, too many "intellectual constructions," as Limerick calls them. The Turner Thesis rests on too many attempts at what rhetoric scholar Ronald H. Carpenter calls Turner's tendency to present "highly stylized" language and his "rhetorical predilection for achieving a strong introductory impact upon his audience" (1983, 36). As a result in the thesis "[w]ords mutate into elastic, freeform mythologies" so that "dissecting the frontier thesis is tantamount to pulling a legend apart" (Jones and Wills 2009, 39).

As such, the Turner Thesis proclaimed the frontier as *the* major component in the "process of self-transformation from corrupted European to perfected American," which had been "central to New World mythology since the seventeenth century" (Madsen 1998, 123). It is a mythological statement, then, that Turner offered, not a historical one. He offered, as David W. Noble put it, "the frontier, the West," as a way to "free Europeans from their ancestral cultures" (2002, 7).

Historians have largely been concerned with questions of Turner's historical accuracy. That is, did Turner present a thesis that could be called historically "True"? But, as the historian Thomas C. Cochran put it in 1970, it is correct that Turner "mainly . . . was right for the wrong reasons, that he was an artist with a new insight, not an analytical social scientist" (1972, 116). But, from a mythological perspective Turner offered less a statement of American history than one of American myth. William H. McNeill described the importance of such statements in his 1985 presidential address to the American Historical Association: "any theory about human life, if widely believed, will alter actual behavior, usually by inducing people to act as if the theory were true" (1986a, 6). The Myth and Symbol scholars were acting as if the Turner Thesis was true and a reasonable description of much of American culture, both before Turner articulated his Thesis and since.

Turner's story of Western expansion in the United States represented a process whereby Europeans with their "Old World [cultural] inheritance," where "suddenly thrust into an infinitely complex environmental and social setting on the raw frontier" (Jacobs 1965, 32). Yet all of Turner's intellectual constructions were exactly that: constructions. Poorly defined as it was, however, the Turner Thesis became an almost Talmudic text in American history for at least three generations. While historians have almost endlessly debated his methods, his rhetoric, and his terminologies, they have continued to use all of them some extent. As one of Turner's more recent supporters, Allan G. Bogue, has put it, while historians actively tried to refute Turner, often "students of Turner's students took up the cause" of the thesis.

In part the reason for this is simple. As Bogue acknowledges in an otherwise spirited defense of Turner, the thesis has "a plasticity . . . that has allowed scholars to use it as a basis for fresh starts" (1994, 215). That plasticity in Turner's ideas mirrored, quite nicely, the plasticity of the "raw frontier" he was attempting to describe. And that kind of plasticity is central to much of Turner's writing—especially those essays that discuss the "significant" of some aspect of the American past. It is those six "Significance" essays, and especially "The Significance of History," that Turner both attempts sweeping historical statements and, importantly, a rhetorical style rich in imagery and poetics (the essays are Turner 1903, 1917, 1925, 1961, 1999).

William Cronon, in a discussion of Turner's "Significance" essays allows for a blunt but still forgiving statement on Turner's intent:

> Like the prophets, he sought not to prove or disprove his vision, but to apply its sweep to all of American history . . . The emphasis on "significance" was a black box which avoided the necessity of more rigorous analysis and theory. It is only when one realizes the essentially hermeneutic nature of Turner's work that one understands why his legacy has been at once so powerful and so problematic.
>
> (1987, 164–165)

It is Turner's attempt at historical hermeneutics that makes his work so applicable to an analysis of American myth. Turner never attempts to account for all available historical data so much as he tries, in Cronon's terms, again, to offer an "essentially *interpretive* thrust [in] his historical projects" (1987, 164; original emphasis). By offering broadly based interpretations of the American past Turner provides a paradigm built on "some of the deepest myths and longings many Americans still feel about their national experience" (Cronon 1987, 160). As a result while more than one hundred years have passed since Turner's initial formulation of his thesis, even detractors must account for it in any attempt to explain American history. While other explanatory "historical models" have come and gone because they are "stopped and inspected at every checkpoint," the Turner Thesis has "operate[d] with a different license entirely" (Limerick 1995, 697–698). That "different license" has been continually renewed, "despite a devastating set of critiques in the 1930s and 1940s," as well as the "full force of Richard Hofstadter's critical intelligence, in his book *The Progressive Historians*, in 1968," that seemed to leave the Turner Thesis "riddled with holes" (Limerick 1995, 698; see Hofstadter 1979, 47–118). Yet, attempts to revise, reformulate, or reject the thesis spend as much time dealing with past historiographical arguments as they do repudiating the thesis, itself, for fighting that thesis, as Patricia Nelson Limerick notes in a particularly honest appraisal, is "like fighting the Pillsbury Dough Boy; it bent momentarily to absorb challenges and then instantly resumed its previous shape" (Limerick 1995, 698).

In large part this was because the Turner Thesis appealed both to the "needs and interests of the rising historical profession," as well as "to the American imagination at large" (Hofstadter 1979, 54). Also of importance was the fact that the Turner Thesis was, as American Studies scholar Alan Trachtenberg described it, "as much an invention of cultural belief as a genuine historical fact." When Trachtenberg goes on to say that the Turner Thesis was "the invention of an America 'connected and unified' in the imagination if nowhere else," he captures both the historiographical intent, but also the mythological result (1982, 17). The Turner Thesis was a statement of American history that accomplished what the American literary historian Van Wyck Brooks would call for in 1918, namely, a "usable past" (1918, 337–341). What Brooks, in detail, suggested was that scholarship in the United States needed to help decide, "*what is important for us?* . . . [what] ought we to elect to remember?" (1918, 340). Brooks may claim to be writing of history, but as will be shown, such questions are far more important mythologically. And, because Turner's essay had so carefully presented American history with that very question in mind, each successive generation of historians' work considering the American West has been forced to deal with Turner, whether through accepting the thesis, revising it, or reacting against it. In each case all historians of the frontier have had no choice but to work through the ideas first presented by Turner. For generations of historians the Turner Thesis will be *the* operating paradigm of American historiography.

From the time of Turner's death in 1932 until the emergence of the new western history of the 1980s the thesis waxed, waned, and was occasionally vilified in American scholarship, but it never disappeared. The thesis was revised, and altered, and expanded, and simplified throughout the century. Historians such as Turner's "champion," Frederick L. Paxson, offered multiple editions of his *History of the American Frontier, 1763–1893* (1967) throughout the century, and Ray Allen Billington offered his own overview of the thesis, claiming that "no one force did more to 'Americanize' the nation's people and institutions than the repeated rebirth of civilization along the Western edge of settlement during the three centuries required to occupy the continent" (1974, 1).

The historical moment where the Thesis moves from strictly an attempt at historiographical explanation to paradigm to full-fledged myth can be found in Walter Webb's 1951 volume *The Great Frontier*. Webb's belief in the validity of the Turner Thesis was so complete, it seems, that he will eventually apply a version of the Turner Thesis to the entire European colonized world and will trace those frontier models back to the emergence of Europe as a colonial power around 1500. In emphasizing the importance of the frontier as a creator and shaper of cultural habits he would contend that the frontier altered institutions, "often in a spectacular manner" worldwide (Webb 1952, 239). Indeed, a large portion of *The Great Frontier* is a Turnerian examination of "the relation between institutions and environment, and their respective influence on the conduct of society" (Webb 1952, 260).

For Webb the frontier provided a unique period in world history in which the frontier—of America, especially, but also in Australia, South Africa, Canada and New Zealand—provided a "great windfall" for the metropole of Western European colonial powers. This windfall, and the associated rise of capitalism, offered a monumental increase in wealth and opportunity of both those on the frontier and those who remained in the colonial power core nations of Europe (Webb 1952, 26–29, 180–202, 239–252, details the rise of this "boom" provided by the frontier for Europeans worldwide since roughly 1500). Drawing exclusively from the American frontier experience and then applying it to all European colonial experiences where "free land" existed, Webb notes that while in Europe the idea of a frontier was "stationary and presumably permanent" and thus represented national borders, in "America it was transient and temporal" (1952, 3). In other words, the frontier represented a massive increase in opportunity for Europeans and represented a unique period in world history, most stereotypically obvious in the experience of Americans on their own, internal frontier.

Bluntly, Webb declared as much a historical statement as a mythical one when he writes "the sudden acquisition of land and other forms of wealth by the people of Europe precipitated a boom on Western civilization, and that . . . boom lasted as long as the frontier was open, a period of four centuries" (1931, 413). For Webb, then, the Turner Thesis could be applied, in a massive totality, around the whole of the globe and could be used to explain, in narrative form, all historical change from the moment Columbus first landed in the Caribbean.

While this is Poor History it Remains Excellent Mythology.

The New Western historians, a movement beginning in the 1980s, would work hard to reject such ideas from the fabric of academic history. The New Western History began as a conference called "Trails: Toward a New Western History" in 1989 (Limerick, Milner, and Rankin 1991, ix). In contrast to the Turnerian frontier "process" New Western historians define the west "primarily as a place—the trans-Mississippi region in the broadest terms, or the region west of the hundredth meridian . . . New Western Historians do see a 'process' at work in the region's history . . . but they reject the term 'frontier' for that process" (Limerick 1991, 85).[5] New Western Historians initiated their program in part as a return to "history" and a rejection not only of the Turner Thesis but also a rejection of the American Studies work on the myths of the frontier. Their interests were formulated by an intent to "widen the range and increase the vitality" of factually based historical research of the "western past" (Limerick 1991, 88). Stephen McVeigh has noted that, if nothing else, what the New Western Historians were interested in was a program in which

> the mythology of the West was being interrogated as it had never been before. No longer was the story of the frontier one of heroism and progress. The New Western historians were producing critical histories that

> saw the frontier as synonymous with conquest, in ethnic, cultural and environmental terms.
>
> (2007, 143)

New Western Historians did not ignore the Myth and Symbol school of American Studies as much as find "the frontier myth" to be "dangerous and destructive" (Thompson 1991, 94). They were far more interested in a New Left version of more traditional historical research. They still understood the "west" or "frontier" had a mythic component and did discuss that aspect of western history. But their interests lay in understanding something that, with a dose of postmodern relativism included, could still be understood as "factual" and "true" about the "real" history of the American West.

This book is less about what is "historically true" than it is about what is mythologically true. Thus, there are no arguments to be had with empirical historians because it is a work more interested in cultural inventions.

As Eric Hayne notes, the west "did not have to be inhabited so much as invented" and as a result its invention still reverberates through American cultural models (1992, 4). It is easy to see the continuing significance of frontier ideology and the Turner Thesis in all areas of American culture. The western genre novels beginning with Owen Wister and developing through Zane Grey, Louis L'Amour, and others are obvious examples. But the romance genre consists of an entire subgenre of "historical romance" that almost exclusively deals with the myths of the American west (Singly 1992 deals with a specific example of this; Kapell and Becker [2005] offers an analysis of the Christian themes of such works). Science fiction, both written and visual, is often defined as a genre of (in the opening phrase of *Star Trek*), "the Final Frontier" linking, in the words of American literary historian David Mogen, "the romance of space travel . . . with the romance of earlier frontier literature" (1993, 16). Even many computer "strategy games" operate through the idea of conquering a frontier, be it on Earth or on the "final frontier" of space (M. Kapell [2002] discusses the Turner Thesis in strategy games).

So, while the New Western historians are no doubt correct that the thesis is either poor or irrelevant history it remains necessary to consider it as a mythic construct. Such a construct is obviously not the realm of academic history.

But, once, it was the central conceit of the field of American Studies.

PARADIGMS LOST: MYTH, SYMBOL, AND AMERICAN STUDIES

As much as Frederick Jackson Turner's Frontier Thesis has been repeatedly attacked as an inaccurate representation of the American past based on "illusions," it has remained an excellent description of American myth. After all, as Jones and Wills put it, the thesis "provided a creation story for the American nation" (2009, 39).

One of the main practitioners of myth and symbol scholarship, Leo Marx, offered a canon of the field in 1969. His canon of classic works of the myth and symbol school includes Smith's *Virgin Land*, R.W.B. Lewis's *The American Adam* (1955), John William Ward's *Andrew Jackson: Symbol for an Age* (1955), and Trachtenberg's *Brooklyn Bridge*; Marx, humbly omits his own *The Machine in the Garden* of 1964 (Lewis 1955; Ward 1955; Marx 1964; Smith 1971; Trachtenberg 1979). Marx's canon is useful but not complete for the project here. As this work deals with various myths of the frontier an urban book like Trachtenberg's has less utility. And, because Charles Sanford's, *The Quest for Paradise: Europe and the American Moral Imagination* (1961) offers versions of the ongoing quest for that frontier myth, it becomes important. Thus, while the practitioners of the Myth and Symbol school would, as Smith himself noted famously in 1957, not call it a unified "method," each of these canonical texts (excepting Trachtenberg) bears some level of responsibility for bringing together images of the Turnerian frontier with concepts of a single, coherent image of an American mythos (Smith 1957).

Although the Myth and Symbol scholars might have been "unorganized" (Marx 1969, 75n1), they were intent on providing a mythic image of America that was unified in the immediate post–World War II era. The new dominant theme in American historiography, which came to be dismissively called "consensus," found Turner's emphasis on regionalism, sectionalism and, in his later works, economics to be far too conflict-oriented.

John Higham coined the term *consensus history* in a 1959 essay for the journal *Commentary* to describe this new kind of history—a term with applicability to Myth and Symbol scholarship as well. Higham described the writing of these new historians to be a profound "change in direction in the exploration of the American past" that was directly reacting to the work of a previous generation of historians of which Turner was a standard bearer (1971, 699). He would later call this an understandably "somewhat conservative view of their country's history" and trace it to a Cold War need to find "an appreciation of [America's] more cohesive and deeply rooted qualities" (Higham 1989, 221). But, when he coined the term *consensus history* it was a form of history presenting a "placid and unexciting past," deadening perceived conflict to make the present appear more homogeneous (Higham 1971, 701). When it came to the legacy of historians like Turner and others of his generation Higham was characteristically blunt: the new consensus historians "feel less need to criticize or defend them than to supersede them" (1971, 700).

Where Turner was interested in conflict—the point between "savagery and civilization," for example—consensus historians wished to, as historian Ellen Fitzpatrick recently described it, "emphasize continuity and stability as the leitmotif of the American past" (2002, 191).[6] The American Studies Myth and Symbol school, too, was primarily interested in the same things, and their presentation of "myth" would be founded specifically on a desire to present exactly "continuity and stability." The postwar historians, whom Gene Wise calls "counter-Progressives" rather than consensus historians, can be described comparatively in this way:

> Where Progressives pictured a society through its *classes*, counter-Progressives saw *culture* as the most basic public reality. Class tends to be divisive, culture more unifying. Sometimes this quest for culture made for a rather homogeneous picture of America.
>
> (1973, 92; original emphasis)

There is an irony here. In searching for the "unifying" aspects of "culture," and in providing histories that offered a "homogenous character," the consensus historians were attempting to accomplish what is, every scholar of myth would claim, the most basic and essential function of myth. And the Myth and Symbol school, in also looking for stability and unity, would provide their version of an American mythos to the same ends.

Sanford's *Quest for Paradise* does this effectively. As John William Ward noted at the time of its publication, "is a book few historians will like [in that it] presents to us now the emotional impact of our national past" (1962, 723). That "emotional impact" required, as far as Sanford was concerned, examinations of Christian notions of Jewish history, and understanding of medieval Europe, a then contemporary understanding of both psychology and myth studies, and an examination of the current political system of the United States. A project of such scope, what Sanford called "something rather different" from the work of Henry Nash Smith, also carried what today are considered major flaws that would make it relatively easy to ignore by most scholars, even at the time (1961, v). Sanford readily admits these flaws and, indeed, did not consider them flaws: "I am concerned," he wrote at the very beginning of the work, "primarily with the larger structure of thought and feeling rather than with textural detail" (1961, vi). That almost unforgivable flaw by contemporary standards pales next to this even greater crime against historiography: "The significance of each is then," he wrote of the topics he planned to cover, "*projected into the present*" (Sanford 1961, v; emphasis added).

In short, Charles Sanford wanted to present an overarching grand narrative of American thought and culture, digging far enough in to the past to offer a significant rewriting of the understanding of the American mythos in the early 1960s. His thesis was more complex than other Myth and Symbol scholars', and like Smith, he concerned himself with the west. However, where Smith looked at the west as virgin land, Sanford saw it as a place to define America against the previous standard of Europe:

> In practice, Americans have consciously or unconsciously sought to establish a midpoint between savagery and civilization, to establish a

> national identity which was neither primitive in the frontier sense nor excessively civilized in the European sense.
>
> (1961, viii)

Where Smith, Lewis, Ward, Marx, and even Turner attempted to discount Europe, Sanford saw not Americans measuring themselves against the frontier but, rather, Americans on that frontier measuring themselves against their own images of Europe. This will be an important perspective in this work as NASA, *Star Trek*, and Gerard K. O'Neill will measure their proposed "new," "final," and "high" frontiers with a similar dichotomous relationship to the past. Where American Studies scholars had argued for the importance of the frontier in understanding American society and history, Sanford went significantly further. The Turner Thesis presented a narrative that was an intentional turn away from Europe, and that turn away, for Sanford, includes many looks over the cultural shoulder, so to speak. Sanford repeatedly sees cultural reactions designed to compare the Edenic New World with the tainted and depraved Old. "The Edenic myth," he claimed, "has been the most powerful and comprehensive organizing force in American culture" (Sanford 1961, vi; see also Sanford 1952). As will be shown, here, the hopes of the "Edenic myth" in Sanford's phrase will be something NASA, *Trek*, and O'Neill will each deal with in its own way.

Yet, unlike other Myth and Symbol scholars, Sanford at least attempts a coherent and detailed theoretical statement of what he means by "myth." In doing so he draws heavily on the work of myth scholars Mircea Eliade and Ernst Cassirer (see, particularly, Sanford 1961, 24–35). Building from the universal myths acknowledged by Eliade, Cassirer, and others of the period, Sanford sees a close relationship between myth and history and builds his thesis from the understanding that "myth also contributes to historical change, for people try to behave in conformity to myth pictures which portray . . . a dramatic world . . . History moves, if it moves at all, in the mass, and mass psychology is peculiarly dependent on myth" (1961, 25). Within that mythic framework, he is able to translate the Puritan "City on a Hill" into Eliade's "Sacred Mountain" and attempts to make the American mythos merely a New World version of a universal—or at least Western—myth system. Where other American Studies scholars will spend an inordinate amount of effort tracing the effects of industrialization, then, Sanford merely needs to note—as he does in both *Quest* and earlier publications—that the "new factory system [emerged] within a limiting, moralistic framework of Americanism which has been called 'the American Myth of the Garden of Eden'" (1958, 2). For Sanford this Edenic myth is neither a small system of meaning nor one unique to America. It is, instead, connected to the entirety of western tradition, and its myths can thus be traced to those sources.

Sandford's focus is overly grand, his conclusions occasionally overly firm, and his method is of its period. But *The Quest for Paradise* is emblematic of the Myth and Symbol school for exactly these reasons. At the same time, however, it moves beyond other Myth and Symbol works. Where the Turner Thesis, and the American Studies scholars who appropriated it, saw Americans as turning away from Europe, Sanford describes a dialectical relationship. But unlike Marx, who saw "the machine" and "the garden" as an internal dialectic, Sanford found one more complex—and more useful in understanding the themes that will emerge in American myth in the late 1960s. "The American, 'this new man,'" is Sanford's topic, like all the American Studies scholars of his period. But his conclusion is that "'this new man' was early conceived in relation to civilized Europe, if not to the savage frontier, as a primitive" (1961, 107). And, for Sanford, the negotiation between images of a decadent Europe, a savage frontier, and some liminal space between them is the ongoing process within American culture.

From one perspective Sanford sees Americans' ongoing relationship with such liminality as one of a people holding "all the materials for a tragic vision of life . . . the classic sin of pride . . . the guilt and shame of failure, the voids of disillusionment." The quest for paradise, which can never result in anything other than a continued movement toward a goal which is forever out of reach, creates little more than "nostalgia for a golden age" (Sanford 1961, 252). As a result Sanford is somewhat less buoyantly idealistic in his conclusions than other American Studies scholars. While the never-ending quest for Eden remains central to American culture, "Sanford finds adherence to the myth a costly affliction and urges instead a more widespread tolerance and—a heightened awareness of life's limitations, a tragic sense of life which will help Americans deal maturely with reality" (M. Fisher 1962, 211–212). At the same time, however, Sanford brings into focus an important theme both of American Studies scholarship and of American culture. *The Quest for Paradise*, although never part of the American Studies canon, serves as an elegant reminder that

> [t]he most hallowed soil in America from its long-time connection with the Edenic dream—where the self-reliant pioneer farmer, driven ever westward in quest of the New Jerusalem, had lost his momentum, become weighted down with mortgages and surrounded by skyscrapers—the Mississippi Valley, "the valley of democracy," became in the hands of America's most talented modern writers a Wasteland.
>
> (Sanford 1961, 255–256; original emphasis)

Finding such an exceptional, new frontier—something other than that "Wasteland"—will be the motive of all the alternate frontier myths in the late 1960s and early 1970s. Sanford offers a version of Myth and Symbol thinking that retains the possibility of looking at more than just the past—it offers the possibility of new dichotomies that might be, using Doty again, "forms of the future." But it offers this just as the critiques against myth and symbol scholarship were beginning to manifest.

REJECTING *POSITIVE* READINGS OF MYTH AND THE BALKANIZATION OF AMERICAN STUDIES

After Sanford any attempt at a value-neutral study of "myths and symbols" will be rejected at precisely the moment in which American society begins experimenting with new derivatives of the traditional mythic themes previously examined by the Myth and Symbol scholars. The main reasons for summarily rejecting the Myth and Symbol school can be placed on the failings of three major ideas in American Studies prior to 1970: myth, holism and "presentism."

To understand how this trinity of myth, holism, and presentism became suspect in the latter 1960s and fully rejected later, requires an analysis of the initial two most stringent attacks on the method as a whole. These are Bruce Kuklick's "Myth and Symbol in American Studies" (1972) and Gene Wise's " 'Paradigm Dramas' in American Studies: A Cultural and Institutional History of the Movement" (1979). Both were published in the major journal of American Studies scholarship, *American Quarterly*, and both attacked the Myth and Symbol movement from specific directions. Kuklick applied the perspective of philosophical meaning analysis to the topics of "mythic symbol" and "presentism," and Wise from the perspective of an insider offering a history of the movement while critiquing the notion of "holism." To the extent that Myth and Symbol school in American Studies had held together because of these three interrelated tendencies, it would be these two complementary critical essays that would lead to suspicion about the conservative, unifying myths celebrated by the Myth and Symbol group. Both essays are a starting point for "The New Americanists" movement that emerged in the 1970s and expanded in the 1980s and rejected the very notion of "myth" as a useful analytic tool at all. And the combined result of Kuklick, Wise, and the New Americanists would be to profoundly misunderstand what "myth" is used for in culture, generally, and how myth can be analyzed and a significant theme in cultures, as well.

Writing in 1972 in what came to be considered the most overtly critical analysis of the Myth and Symbol school of American Studies, Bruce Kuklick called the publication of Henry Nash Smith's *Virgin Land* a "major intellectual event" (1972, 435). He then proceeded to disassemble Smith's work—and with it, all of the Myth and Symbol school—along negative philosophical lines. Since 1970 when Robert Sklar declared "American Studies, as an intellectual discipline, is in crisis," scholars began arguing for the superiority of contemporary work. Any use of myth or frontiers became exclusively used to argue for the intellectual bankruptcy of previous models (Sklar 1970, 597).

For Kuklick, a far more traditional historian than most members of the American Studies movement, the "crisis" in American Studies scholarship was based, first, on a fundamental philosophical issue and, second, on historical methodology. The philosophical issue revolved around the American Studies use of "myth" and the historical methodological issue the attempt to make those myths relevant to contemporary audiences—or, in Kuklick's term, "presentism." Combined, these two significant problems in the work of American Studies scholarship negates the field's work as a whole and throws into doubt the utility of future work. As we shall see, Kuklick's dislike of the Myth and Symbol school is based, in the final analysis, on a dislike for the lack of empiricism in "myth" as a category, and the study of "myth" as a method. In other words, Kuklick's issue is less with the Myth and Symbol school than with the concept of "myth," itself.

Philosophically, each of the major works in the field contained, "a set of classic errors" based on a mistaken Cartesian assumption about the relationship between categories of "myth and image" and "empirical fact" (Kuklick 1972, 450, 438). The main conceptual failure for Kuklick hinged on the idea that American Studies produced a definition in which "symbols and myths at best *reflect* empirical fact, and so are never themselves factual" (1972, 436). Moving from a worry over the lack of empirical standards, Kuklick's other issue of concern is that these nonempirical constructions are then used to draw broad conclusions. "One of the primary purposes of the American Studies movement," Kuklick notes, "is to demonstrate the way in which these 'collective' images and symbols can be used to explain the behavior of people in the United States" (1972, 437). Kuklick is undoubtedly correct to note the problems faced in the attempt "to relate intellectual currents to the culture's zeitgeist and to argue that some symbols and myths dominated all America" (Kuklick 1972, 443). One of his major claims is that Myth and Symbol scholars had not offered enough empirical evidence to generalize from the texts they offer to American culture as a whole. As Jeffrey Louis Decker noted in his critique of Kuklick, his "constant recourse to the importance of studying 'American behavior'" remains his main argument against the way scholars such as Smith and Marx assess cultural "truth" (1992, 296).[7] He is also correct in describing the difficulty Myth and Symbol scholars had in supposing that "their myths and symbols form a hierarchical structure which has a consistent and verifiable relation to those more specific beliefs" of the culture at large (Kuklick 1972, 446).

Furthermore, the desire among Myth and Symbol scholars to have their work offer meaningful insight not just into the American past but into the present culture as well is, in Kuklick's dismissive term, "presentism." For Kuklick, this is "notorious among the errors that historians can make—interpreting the past in concepts applicable only to the present" (1972, 441). Continuing his critique Kuklick depicts presentism as a series of Platonic forms that permit the historian to examine "a set of eternal ideas existing independently of the individuals thinking about them." The historian's job becomes one of the categorical comparisons. "It is, therefore, easy," Kuklick goes on, for a "historian to formalize his present concerns . . . and then read these conceptions into the past" (1972, 441). This is precisely the mistake made by American Studies scholars in Kuklick's critique. They have, as historian David Hackett Fischer put it, provided a "retrospective symmetry" in their attempt to provide a narrative that is useful to present-day readers (1970, 140).

Kuklick's critique was the critique of a historian, and specifically an historian applying an empirical meaning criterion derived from analytic philosophy. While American Studies, as an interdisciplinary field, used both the primary sources used by historians, and in many cases the methods of historians, the work remains very much *not* the work of historians. What Kuklick considers to be the most crippling of flaws in American Studies research are actually requisites for mythological and mythographical work—*and for myth, itself*. Yet, for Americanists, the end result of Kuklick's detailed and careful analysis of what he argued to be a somewhat specific failing on the part of Myth and Symbol scholars was that the idea of examining something that could be called "myth" in a generally positive sense became anathema. Previous scholarship came to be seen as "idealized and romanticized accounts of the evolution of a white patriarchal America" (Paul 2014, 21). A new generation of scholars began using myths of the frontier for avowedly negative readings of both the American past and present.

Kuklick remains apologetic but firm in what he calls this "philosophical critique" of the "humanist methods" (1972, 447) of the Myth and Symbol school. At the same time, however, within that apology comes a paradigmatic bias in favor of a more traditional form of historiography. Kuklick remains convinced that at issue with a myth and aymbol analysis is its lack of empirical data. He uses terms like "empirical data," and "empirical procedure," nearly as frequently as "Myth and Symbol." He then ends his essay with some contrition, pointing out, "I realize that philosophical criticism is much easier to do than constructive empirical research" (Kuklick 1972, 450). What Kuklick argued for, explicitly, was a return to more stringent historical methodology. It was not the "Myth and Symbol" school that his philosophical critique was directed against: it was anything as "fuzzy"—and, thus, not "empirical"—as the myths and symbols, themselves. Because what myths and symbols amount to, for Kuklick, "are ruminations with little empirical content, and not history" (1972, 447).

Even in defense of American Studies most scholars have accepted Kuklick's terminology and classification, although Trachtenberg would later note that the implied "coherence" of the Myth and Symbol school was "more an effect of the attack launched against it than of any collective program or agenda" (1984, 668). That attack was both intense and long lasting.

As concern the "holistic" aspects Robert E. Spiller had already offered in 1960 insight into what that meant when it came to the method used by American Studies scholars. Referring to the work of Henry Nash Smith,

Spiller noted that Smith's conception of American Studies offered "a concern for an increased understanding of American culture" (1960, 213). On the surface this seemingly positive remark carries with it some rather significant landmines. Spiller quickly delineated those landmines. "The existence of such a culture is assumed," he noted, but for American Studies academics that assumption carried a principle that

> [i]f culture is defined as the way in which subjective experience is organized, there remains only to discover who has done the organizing, exactly what components have been organized into a single whole, and how to study the result . . . the tacit assumption [is] that a single method . . . would be desirable for research in American Studies [because of a] desire to find an organic unity in the subject itself.
>
> (Spiller 1960, 213)

That "organic unity in the subject" is the implicit assumption that Gene Wise identifies as a major theoretical issue for American Studies. Unlike sociologists or anthropologists who have defined holism as the necessary methodological approach for understanding the whole of a group by the aggregation of somewhat arbitrary cultural aspects that can be applied to the group as a whole, American Studies operates from the assumption that these cultural aspects are part of an organic reality that *do* make up the entirety of the American culture. For Wise this theoretical problem reads somewhat more specifically, however. "There is an 'American Mind,'" Wise declared as a definition of the holistic assumption of the field and, that "mind is more or less homogeneous. Though it may prove to be complex and constructed of many different layers, it is in fact a single entity" (1979, 306).

This is a far more significant argument than Kuklick's. Wise's critique comes with the implicit assumption that American Studies had simplified American culture through a form of homogenizing reductionism. As Cecil F. Tate put it, the assumption among American Studies scholars about holism also strongly overlapped the critique made by Kuklick when discussing his view of presentism. Tate described holism this way:

> Holism is first of all a belief about the nature of culture. Traditionally it means simply that there are such things as identifiable cultural units. The "unit" is not a mere abstraction, but something which exists prior to, and independently of, the cognitive activity of the human mind. It simply is as a part of reality.
>
> (1973, 11)

As Kuklick had noted, American Studies scholars often merely described the past in the form of these "units" and called the result presentism, Wise notes that this approach also allowed for the belief on the part of scholars that their purpose was to holistically "probe for *the fundamental meaning of America*" because " 'thought' in America is an integrated whole" (1979, 307; original emphasis).

Wise's critique differs from that of Kuklick because Wise worries far less about the inherent empirical flaws of an American Studies method. At the same time, his critique remains a strong one: the canonized texts that make up the Myth and Symbol school do all retain definitions of "myth" and "culture"—and, specifically "American culture"—that are highly problematic. At the same time, however, these issues resolve themselves with a more contemporary version of "myth," which has been offered here. However, at this juncture it is possible to use Wise's own words as a strong indicator of what this study attempts to undertake with regard to the myths of the frontier inherent in the latter 1960s. In his closing paragraphs Wise notes about the American Studies movement precisely those aspects that this study attempts to uncover in the differing experiences of different American groups with regards to NASA and Vietnam and *Star Trek* and Gerard K. O'Neill's High Frontier:

> We have moved beyond the block assumption that there is a single holistic "American Culture," expressed in "The American Mind," to a more discriminating consciousness that contemporary cultures function on several different levels, and in several different ways. We are less inclined now to take readings from a single vantage point on *The* American Experience; instead, we look upon America from a variety of different, often competing, perspectives.
>
> (1979, 319; original emphasis)

As a result of these critiques, by the end of the 1970s there seemed to be no possibility of adjusting the Myth and Symbol school's methodology for new analyses based on myth. Ann Fabian, in her later retrospective review of *Virgin Land* and other books of the Myth and Symbol school described her experience as a graduate student reading them for seminars in the period:

> We dismissed them all. We noted their oversights, their lack of theoretical rigor, their simple confidence that the varied experiences of past actors could be captured by a single unifying myth, and their inattention to the necessary facts of the past.
>
> (1996, 542)

Fabian's description of the rejection of the Myth and Symbol school explains how future mythographic projects were organized around wholly negative valences and explains why the New Americanists in American Studies could so easily ignore earlier Myth and Symbol scholarship in their field.

Like the New Western historians' attempt to "lay siege" to the Turner Thesis in the 1980s, American Studies also had a new generation of scholars, educated like Ann Fabian, who felt the need to make permanent the sharp institutional break that immediately preceded them. For a time after the early 1970s into the 1980s, scholarship took the route of overtly critical and negative theses of frontier myths; then such topics were jettisoned, and any Myth and Symbol work became truly verboten. Later, myth work was done, of course, but largely ignored by American Studies practitioners. Where the Myth and Symbol school received their name through critique—as did the consensus historians—these scholars declared themselves, in a fit of self-assurance, "The New Americanists." Emerging also in the 1980s, but accelerating only in the 1990s and after, the New Americanists firmly divorced themselves from their institutional history. Taking just two edited volumes as an example of the New Americanists, Donald E. Pease and Robyn Wiegman's 2002 *The Futures of American Studies* and Philip Fisher's 1991 *The New American Studies*, it becomes clear that the movement is defined almost exclusively from American university departments of literature and English (P. Fisher 1991b; Pease and Wiegman 2002b). Fisher's volume, the contributions of which were taken from the critical theory journal *Representations*, has thirteen contributors, twelve from departments of literature or English. Pease and Wiegman bring together twenty-three contributors, eighteen from literature or English programs. None of the contributors of either volume is a historian—at least as determined by being employed in a history department. There are no sociologists, a single political scientist, and the occasional senior scholar appointed in "humanities" generally. And, for a field so concerned with "culture," it is shocking that neither has a single anthropologist. It is clear that the interdisciplinary tradition of American Studies, for the New Americanists, can be housed in only one kind of department. And, not surprisingly perhaps, as a result the focus tends to move decisively away from the study of anything other than literature and rhetoric.

Philip Fisher, in his introduction to *The New American Studies*, delineates this movement quite pointedly and in ways that nicely capture the ideology of the New Americanists, though they would deny any homogeneous ideology whatsoever. "This new generation of American studies" Fisher declares, "has passed from myth to rhetorics." He is correct, but would be more correct had he openly declared American Studies had passed from interdisciplinarity to find a home in his own department of English. This shift, which is not surprising considering the earlier critiques of the use of myth in American Studies, is for a simple reason based on a poor definition of myth: "myth is fixed, satisfying, and a stable story used again and again to normalize our account of social life . . . *myth* in this perhaps too simple formula is always singular, *rhetorics* always plural" (P. Fisher 1991a, vii; original emphasis). For Fisher, who is quite representative of all the New Americanists, "Rhetorics are also distinct from ideology" (1991a, vii). Lest the point be missed, Fisher quickly lists the accepted canon of early American Studies myth and symbol works—including most of those discussed here—and notes that the purpose of those works, and myth and symbol scholars in general, was to undertake "the search for a central myth of America" (1991a, viii). For Fisher, however, the study of such myths was to always result in "the disappointment of myth by fact" (1991b, ix–x). For the New Americanists—as with so many other scholars at this point—the "problem" with myth laid in the assumptions of empiricism. This kind of critique, however, brings with it its own set of problems. As William James described it earlier in discussing his form of "truth," it is a critique that means the "desire for a certain kind of truth . . . *brings about that special truth's existence*" (1960, 24; emphasis added). In rejecting the category of myth the New Americanists are creating their own myth of empirical truth for their form of scholarship.

It only works if they believe it enough.

If Myth and Symbol scholars were indeed looking for Fisher's declared "central myth of America"—which they certainly were—then the New American Studies was defined not by central myths derived from "consensus," but by fragmentation of American culture defined by the Canadian Americanist Sacvan Bercovitch as its opposite: "dissensus" (1986a). Fisher, as is the case with the majority of New Americanists, is interested in those things the original members of American Studies missed. So, while Myth and Symbol scholars were interested in defining myths of a unified culture, New Americanists are interested in those aspects that drive people apart. But, as Sacvan Bercovitch had already noticed in describing this trend, the examination of those things Americanists disliked about the myth and symbol paradigm were driven by their own personal definition of that dissent. As much as the Myth and Symbol scholars were an organic extension of the consensus history of the same period, Bercovitch noticed, the dissent came from those who experienced the late 1960s. As he put it in describing the contributors to his *Cambridge History of American Literature*, but in a way easily applicable to the New Americanists in general: "To start with, we decided to go with those whose dissensus it is; that is, with Americanists trained in the sixties and early seventies" (Bercovitch 1986b, 634).

The dissensus scholars, unlike the consensus scholars of a previous generation, were interested in undermining hegemonic master narratives. The Myth and Symbol school, driven by the post–World War II ideological project of consensus, was precisely that. Thus, the critiques of the Myth and Symbol school were predicated on the fact that those mythographic analyses were unwittingly ideologically driven—what Leo Marx would ironically describe as the "nationalistic, patriarchal, racialist, hegemonic 'master narrative' "—perpetrated by Myth and Symbol scholars prior to the end of the 1960s (2005, 123). Thus, it makes perfect sense that Fisher's claim, that the study of a plurality of "rhetorics" has no ideological implication for the scholars involved but is merely the study of "what is uncertain or potential in culture" (1991b, vii).

Yet, as a final result, any topic that might include the term *myth* was to be avoided at all costs, lest one drift into ideological presumptions, even as newer understandings of the term have come to include exactly what Fisher is searching for, the "potential in a culture." "Ideology," the New Americanists claimed, functioned as the Myth and Symbol scholars had seen "myth," that is, "a cultural mechanism of stabilization and transmission, neither of which is a primary topic of a culture of speculation" (P. Fisher 1991a, xxii). Those scholars who did examine myth did so to note Marx's "nationalistic, patriarchal, racialist, hegemonic 'master narrative' "—that is, they examined myth from the perspective of what could be found with primarily negative consequence. As a result, with Kuklick's and Wise's early attacks on the Myth and Symbol school, and with the rise of the New Americanists, the fundamental purpose of American Studies was lost. It is not a surprise that so many New Americanists find themselves all in the same kinds of literature or English departments. That earlier generation of American Studies scholars they have rejected, after all, had "claimed the power to integrate their incompatible knowledges within a single interdisciplinary agenda" (Pease and Wiegman 2002a, 16). The critiques of Kuklick, Wise, and the New Americanists are all correct in a limited fashion, and what American Studies had done in its "classic" era was, as Sacvan Bercovitch put it, to reveal an "America" that "appeared out of nowhere" (1991, 979). But the result of these rather specific early critiques of American Studies methodology have been the pulling away from any topic of too large a cultural dimension, and the narrowing of focus to smaller studies comfortable in the confines of academic departments and planned to be read by the populations of other academic departments and, if truly successful, the students in those departments as well.

But ignoring myths, or finding only negativity in them, meant the ongoing critiques of "myth" in American Studies failed to notice that the study of mythography had, like American Studies, American history, and other fields, also evolved. The straw man current American Studies scholars are rejecting is exactly that—a straw man, built in the middle of the last century and easily vanquished as a result.

A NEWER APPROACH TO MYTH

To examine precisely why this American Studies version of "holism" failed to account for mythographic changes in the latter 1960s requires a more comprehensive understanding of "myth" in today's scholarship. Perhaps more categorically, however, the borrowing of both schools of "myth" and "holism" from functionalist anthropology contained a specific intent as well. Malinowski, like all anthropologists of the period, intended their work to be, explicitly, an empirical science. As British anthropologist Edmund Leach described Malinowski, "his first line of evidence was always first-hand observation by the field-worker himself. Empiricism could hardly be carried further" (1957, 120). Indeed, Malinowski would go to great lengths to call his method of research "ethnographic empiricism" (1923, 221). But long before Leach described Malinowski's method as, at best, "searching all the time for concepts of the middle range of generality . . . not so abstract as to amount to mere verbal speculation, not so concrete as to defy generalized comparison" and thus not particularly "empirical" at all, the question of an empirical anthropology had already been answered in the negative (1957, 136). Yet, it would be the quest for an empirical form of Myth and Symbol that would eventually undo the first generation of American Studies scholarship. That generation, after all, was attempting to combine an empirical study of myth in combination with the extended scholarship of the frontier initiated by Frederick Jackson Turner. And, while Malinowski's empirical anthropology failed to be, strictly, empirical, Turner's work suffered the same problem. Turner's career, as the American frontier and environmental historian William Cronon described it, "never managed to combine the rigor of his scientific methods with the synthesis of his oratorical rhetoric" and, like Malinowski's scholarship, ended with "concepts of the middle range of generality" (1991, 93).

Even when American Studies scholars were busy adapting functionalist approaches to myth from anthropology and combining it with the Turner Thesis, the anthropologists themselves had already realized that such work was not, by any definition, empirical, and the historians had long since given up on granting Turner the status of social scientist. Yet the connection American Studies scholars were aiming for with their attempt at an empirical study of American myth was one that took empirical truth as the route to mythical ends. As Fred Welbourn classically stated in another context about "myth, rationalism and empiricism" in 1964, American Studies scholars presented a mythographic method that seemed to believe if it "pursues [its] empiricism far enough, and applies its conclusions with thoroughness, all problems will be eliminated" (1964, 41). This intended conclusion that was, finally, more of a belief about myth perpetrated by American Studies scholars than anything approaching an empirical supposition, tinted all Myth and Symbol scholarship in the period leading up to the late 1960s.

As a result the critiques of the Myth and Symbol school attack two things: the implied version of "myth" within that scholarship and the empirical status of both "myth" and the "frontier" as a cohesive idea as well. The attack on "myth" is an attack on an especially conservative understanding of the term but has been quickly accepted as an attack against "myth" in general. And, while it is true that the Turner Thesis is, from a historiographical perspective, riddled with flaws, this also does not preclude its application to an understanding of contemporary American culture. The critiques and attacks against the Myth and Symbol school are all, in the final analysis, attacks based on the perception that this particular "myth of the frontier" perpetrated in classic works cannot, in any real sense, be accepted as *empirically true* and where they do affect contemporary society do so only in a negative way. At the same time, in an attempt to appear empirical Myth and Symbol scholars slipped into the erroneous method of reductionism about what "myth" is and how it works. This trap—or perhaps this hope, as Richard Rorty put it in 1963—that it was possible to be "*bona fide* empiricists without being 'reductionists,'" is a difficult one to avoid (176). But in attempting to avoid the trap, or believe the hope for consensus, American Studies Myth and Symbol scholars did become reductionist, and also opened up their work to narrowly focused but significant critiques for which they had limited defenses.

The attack on the Myth and Symbol school of American Studies is an attack predicated on few issues. But, it is an attack based on the correct premise that the Turner Thesis is flawed—that it is not good history—combined with a poor definition of myth—that it is simple, static, and conservative. This does not, however, prove that the conclusion that the idea of a mythic significance of the frontier is thus incorrect. This is a fallacy built on the idea that since the argument for a conclusion is false the conclusion, itself, must be false.

This book argues that, while Turner can be seen as an influential but flawed historian his argument for the significance of a frontier is, in fact, excellent—if unintentional—cultural mythography.

FRONTIERS AFTER VIETNAM

If frontier myths were to be discussed at all after the period this book examines, they would be done so almost exclusively from a negative perspective. The most obvious examples of this trend are Richard Slotkin's works on the myth of the frontier that begin with his 1973 volume *Regeneration Through Violence* and continues through the 1985 *The Fatal Environment* and the 1992 *Gunfighter Nation*, and Annette Kolodny's *The Lay of the Land* of 1975 which saw earlier forms of a Virgin Land as "gendered symbolism [and] a metaphor of rape and patriarchal exploitation" (Slotkin 1973, 1993, 1998; Kolodny 1984; Paul 2014, 21). Slotkin's understanding of the frontier myth takes Turner to task for his presentation of a mythological structure already firmly in place at the time of his 1893 presentation of the significance of the frontier. This mythological structure that is reified by Turner is one in which "the warrior hero, and the violent means by which he has gained . . . wisdom, are inseparable," and the violent action found in the three hundred years of literature and history Slotkin discusses, cause the hero to "initiate" a "regenerative process" (1998, 374–375). For Slotkin the mistaken idea of American Exceptionalism is understood through a literary, historical, and mythological construct of frontier heroism that continues to color American culture up to the present. It is a process, as reinforced by Turner, in which conflict and violence are central. From its very beginning "Europeans had to struggle against and unfamiliar environment and against the non-European, non-White natives" (Slotkin 1993, 11).

Slotkin calls this process, in his initial conception of it, "regeneration through violence" in which the New World is, indeed, made "new"—is regenerated—through violence. This is the Turnerian conception of the frontier's erasure of European cultural models that are rebuilt as fully American. For Slotkin there are profound psychological underpinnings of this mythological superstructure. It is a mythological struggle in which white, "male avatars of wilderness and civilization" struggle for "possession of the white female captive—a figure who embodies the Christian moral and social law that the hunter both defends and tries to avoid" (Slotkin 1973/2000, 179). It is, then, a violent cultural contest between savagery and civilization.

If these negative implications of how Slotkin presents his version of the "myth of the frontier" are not convincing enough, the somewhat more obvious nature of the work is made plain by his references to the Vietnam War. In his three volumes, which are magisterial in scope and scholarship, Slotkin manages to bring in references to Vietnam regardless of his period. Vietnam certainly fits within the scope of the final volume, which deals with the twentieth century. Yet even in the first volume, published in the early 1970s while fighting in Vietnam was still included American soldiers, and covers the period from 1600 to 1860, sees mention of the war. Slotkin manages to note that the most recent use of the mythology he is discussing "has been on behalf of escalating the war in Vietnam" (1973/2000, 562). By his volume on the period from 1800 to 1890, published in the middle of the Reagan presidency, Vietnam is not merely an aside in the text but is a central part of his introduction. He quotes a Vietnam Veteran saying, "The Indian idea . . . the only good gook is a dead gook . . . Some people were on an Indian trip over there" (1998, 17). Slotkin's purpose is to expose the prevalence of such myths and the rhetoric that supports them, but a decade after the end of both the Myth and Symbol school and the Vietnam War, such mentions can also be read as reactive to the present moment in American society.

This is not an attempt to devalue the work of Slotkin—or any other scholar after the demise of the Myth and Symbol school that still approached the topic of "myth" in any way—but to note their conception of frontier myths had been heavily influenced by the Vietnam War. Indeed, Slotkin's three volumes are far too multifaceted and extensive to belittle with the phrase "Vietnam reactionary texts." But, in light of the cultural reaction to Vietnam it is not surprising, either, that Slotkin both manages to note the war in each of his three volumes and that his reading of a "frontier myth," cultured by the American experience in Vietnam, becomes very much one of "frontier violence."

MYTHS AND SYMBOLS OF THE LATTER 1960S

What this genealogy means for this work is actually quite precise. The rejection of the Myth and Symbol school of American Studies, and the continual debates and eventual rejection of any descendent of the Turner Thesis, when combined with an ongoing theme of both empirical research and multicultural realities has meant that there is no possible way within contemporary academe to offer a synthesis of the frontier myths being explored by American culture at the end of the 1960s and the beginning of the 1970s. Yet, as this work will highlight, certain aspects of myth and symbol scholarship can offer useful insights into the period. From the "Virgin Land" hopes of NASA's lunar landings, *Star Trek*'s voyages that repeatedly find paradisiacal planets, and Gerard K. O'Neill's utopian High Frontier, myth and symbol scholarship provides motifs to build on. Smith's *Virgin Land* is encountered often, and in Sanford's *Quest for Paradise* an impetus for each of these topics can also be found as well. Ward's reduction of those myriad symbols into the person of Andrew Jackson offers striking opportunity to examine the characters of *Star Trek* and the life story of Gerard K. O'Neill. And, as O'Neill offered up a new way for new Adams to move into the future and *Star Trek*'s United Federation of Planets provided new Eden-like fictional locales for those new Adams to journey toward, each of these works of American Studies scholarship retains utility. When a fully current understanding of what contemporary first class scholars of myth mean when they examine their topic, a deeper and polysemous understanding of the many American "myths of the frontier" can be also be considered.

The debates surrounding, first, the Turner Thesis and eventually the Myth and Symbol school were often concerned with the efficacy of the terms employed, be they "frontier" or "myth." Too often, critics eventually realized, those terms attempted to carry too much meaning with too little definition to meet empirical standards. The collapse of any coherent positive study of American mythologies in the late 1960s was based on these rejections, and was decried because both the "frontier" and the "myths" of that frontier attempted to encompass too much. It might be said in the same way that a similar problem manifests itself, then, in discounting those critiques with yet another single term. But that term—much like the others—offers a metonym for a range of changes in the period. Much like the consensus historians were unable to explain the single most important event of the nineteenth century, myth and symbol scholars were presented with a similar event that seemed to preclude even a vaguely positive reading of the "myth of the frontier." Additionally, like *frontier* and *myth*, it is a term that has come to carry the weight of far too many themes in recent American history and seemed to undermine any possible pretense that the myths that had fueled the American past could be of use in the present, even should the hopes of Doty's "forms of the future" be ignored.

And that term is, of course, *Vietnam*.

The rejection of myth by the New Americanists was the rejection of an idea of myth dating to the middle of the last century after all. And the "frontier ideology" inherent in the war was an ideology that encapsulated the very worst of previous frontier myths' racism, sexism, colonialism, and a host of other ills. So to discuss how the Vietnam War undermined any positive notion of American myths, it is necessary, first, to offer the current version of myth.

PARADIGMS REGAINED: MYTH IS A POSTMODERN WORLD

Necessarily, then, this work must present a more contemporary version of an understanding of the term *myth*. There is an irony that myth and symbol scholarship was rejected by American Studies scholars for its lack of empiricism at the precise time so many frontier myths were emerging in the United States and postmodernism was moving to dominate thought in American academia. As Percy Cohen noted in 1969, myths are quite simply *not* "attempts to make correct empirical statements" (1969, 353n1). The various attacks on the Myth and Symbol school are actually a rejection of "myth" as a topic worthy of intellectual analysis in a positive fashion by a generation of scholars who hoped for—and in the end required—a claim to empiricism in scholarship. Instead, myth became a negative, static, and unchanging construct—not one capable of major cultural shifts.

A major problem with employing the term *myth* in this work is that scholars of myth are notoriously incapable of defining the term. In many ways scholars all agree that they know what "myth" is about, but when they approach the theoretical details of agreeing on a definition the arguments begin. American literary scholar of science fiction Frank McConnell understood this difficulty explicitly, saying that "of course the moment you think you know what a myth *means* you have lost contact with the myth itself" (1979, 285; original emphasis). William G. Doty has confronted this issue in detail. For Doty the definitions fall into a wide spectrum, "from 'outright lie' to 'most important underlying source of cultural life'" (2004, 11). As a result scholarship about myth by scholars in American Studies and related disciplines follow what may seem a peculiar or eclectic path because, as Doty has noted, "myths can be made to mean whatever the myth teller wants them to mean, and their rhetorical power can be subjected to the prevailing modes of discourse of a particular era or power elite" (2000, 25). This is, it should be noted, precisely the critique leveled against myth and symbol scholarship since the 1970s and remains an issue this work attempts to avoid.

But, in addition to this, any attempt to define *myth* must account for a complex combination of issues. Of main concern are three things: the *structure* of myths, the *function* of myths, and, directly relating to the first two, the narrative organization *of* myths.[8] This has, after countless attempts by many myth scholars over the course of literally centuries, proven to be an impossible task, and usually fruitless as well. Each attempt at a rigorous universal definition of myth has proved to be a failure because such an attempt defies at least some of the very reasons for the existence of myth. Definitions that satisfy structural requirements tend to fail functionally, and vice versa. Definitions that tend to describe narrative universals of myths tend to fail either structurally or functionally, if not both. But, as shall become apparent, it is the act of attempting a definition, itself, which is most problematic. Such definitions, bound up as they are in empirical notions of precision, fail because myth cannot, and should not, be so rigorously defined. To do so undermines the very purpose and power of myth, itself.

As a result it is useful to begin from a more basic set of assumptions and turn to Mark Schorer's well-known statement that nicely sidesteps this problem by offering a very general definition:

> Myths are the instruments by which we continually struggle to make our experience intelligible to ourselves. A myth is a large, controlling image that gives philosophical meaning to the facts of ordinary life; that is, which has organizing value for experience.
>
> (1968, 355)

This definition, it should be noticed, captures much of what the myth and symbol scholars attempt in their various works discussed here. Indeed, it is possible to simplify Schorer's definition even more. American Studies myth and symbol scholars would argue that a very general definition of myth, as proposed by Bruce Lincoln, that myth is, in the most simple form possible, an "ideology in narrative form" (Lincoln 1999, 207).

Lincoln's definition of *myth* is overly general, but it works. It works because previous attempts to define the term in greater detail have failed. As will become evident, a definition, by its very nature, cannot be part of a mythological view. British classicist Geoffrey S. Kirk put this quite clearly, stating,

> There is no one definition of myth, no Platonic form of a myth against which all actual instances can be measured. Myths . . . differ enormously in their morphology and their social function.
>
> (1970, 7)

It is this lack of a single and simple definition that remains the problem among nonmyth specialists. In order to more precisely get to this issue it is necessary to fully articulate not what myth is, but how mythic discourse departs from the rules usually applied to academic scholarship. Those rules, which are the central plank in many of the critiques emerging against the Myth and Symbol school in the 1970s, are based primarily on empirical evidence and logical consistency. To put that into the terminology of myth scholars, that is the rules are based on *logos*. Logos, a Greek term usually paired as the opposite of myth, is best described by Karen Armstrong. If we extend what Armstrong relates about logos to much of modern academia, and pointedly to American Studies, the term is quite enlightening for the whole debate on the Myth and Symbol school of scholarship. For Armstrong, logos

> must relate exactly to facts and correspond to external realities if it is to be effective. It must work efficiently in the mundane world. We use

> this logical, discursive reasoning when we have to make things happen [or] get something done . . . *Logos* is practical. Unlike myth, which looks back to the beginnings and to the foundations, *logos* forges ahead and tries to find something new; to elaborate on old insights, achieve a greater control over our environment, discover something fresh, and invent something novel.
>
> (2000, xiv–xv)

It is not a particularly long trek to note that the many critiques directed against the Myth and Symbol school of American Studies are, in the final analysis, predicated on the concept of logos. Myth and Symbol scholars aim for a mythographic analysis, but their critics will all desire empirical proof of a "correspondence to external realities." As the terms are so nonspecific in myth scholarship, the empiricism is always left wanting. To contrast those critiques to what Myth and Symbol scholars have attempted it is useful to look to Karen Armstrong's contrasting definition of myth. For Armstrong, in the past at least, myth

> was regarded as primary; it was concerned with what was thought to be timeless and constant in our existence. Myth looked back to the origins of life, to the foundations of culture, to the deepest levels of the human mind. Myth was not concerned with practical matters, but with meaning . . . the *mythos* of a society provided people with a context that made sense of their day-to-day lives; it directed their attention to the eternal and the universal.
>
> (2000, xiii)

This difference between the empirical-minded logos-centricity of the critiques of the Myth and Symbol school and the less "practical matters"–oriented version of myth scholarship being advanced here bears some scrutiny. It is of primary importance that a comparison between the two—between the mythos and the logos—not be made too explicit. One single line from Henry Nash Smith's classic Myth and Symbol work *Virgin Land* has, over the years, received intense criticism, but it is one that applies here as well. Concerning "Myth and Symbol," Smith claimed, "I do not mean to raise the question whether such products of the imagination accurately reflect empirical fact. They exist on a different plane" (1971, xi). Many critics of both Smith's book, and the Myth and Symbol school as a whole, took that phrase as proof of the problems with the method. But, regarding a comparison between logos and mythos it is necessary to maintain that they do, in fact, "exist on a different plane" and any comparison is doomed to marginalize one or the other. Mythos and logos are not best described, as Heike Paul has put it, by "a critical perspective on myths by contrasting them with 'truth' ('logos') or 'scientific thought;' myth here considered false, fictional, anachronistic, 'primitive,' or 'pathological' " (Paul 2014, 26). They are, instead, two very different ways of understanding the world. Furthermore, even should we use a contrasting definition the assumption that mythos would be the marginalized part of such a comparison does not necessarily follow.

Thus, while this work aims to ignore, as much as possible, a direct comparison between mythos and logos, it does very much agree with the philosopher Lawrence J. Hatab on the subject:

> The point is not to protect myth from various forms of rational-ism [*sic*] but also to acknowledge the links between myth and reason, which can go a long way toward resolving perennial philosophical problems which had been created by rationalistic, metaphysical, and reductionistic assumptions in the first place.
>
> (1990, 326)

It should be somewhat clearer by now that the real thrust of all the critiques against the Myth and Symbol approach in American Studies was not about "method" so much as it was about differing conceptions of what scholars mean by "truth." It is only necessary to note that, for the critics of the Myth and Symbol school, an empirical form of "truth" was the intended purpose of all scholarship, and scholars who ignored such designations should have their work rejected. As British historian Stephen Davies rather sharply puts it, empirical history is an attempt to supplant "myth, tradition and epic" because empirical history can lead to the question "is it true?" and mythology cannot (2003, 126). Rather than worry overly about quanta of empirical evidence however, this study accepts Karen Armstrong's assertion about myth:

> Myth, therefore, *is true because it is effective*, not because it gives us factual information . . . if it works, that is, if it forces us to change our minds and hearts, offers us new hope, and compels us to live more fully, *it is a valid myth*. Mythology will only transform us if we follow its directives.
>
> (2005, 10 emphasis added)

While Armstrong's rhetoric is somewhat expansive, she does capture a necessary statement about "myth" and "truth" together. As such, it is important to offer, briefly, an exact meaning of how "myth" can be considered "truthful" in a nonempirical sense.

A PRAGMATIC FORM OF TRUTH, A PRAGMATIC FORM OF MYTH

Recent scholarship in American Studies—especially among those who classify themselves as the "New Americanists"—has taken some delight in noting that traditional scholarship in the field has had a "nativist" tendency. In their willingness to use various schools of poststructuralist and postmodernist thought the New Americanists have resisted drawing from previous "American" traditions. However, in presenting a notion of "mythic truth" the only academic definition of "truth" that rightfully applies is that which comes from the American school of pragmatic philosophy. While any contemporary New Americanist will rightly point out that pragmatism is, at best, a philosophical approach with a substantial number on non-American precursors, in the sense of conceptualizing a kind of "mythical truth" for this work ideas deriving from the pragmatism of William James applies.[9]

To understand a fully pragmatic concept of "mythical truth," however, it is necessary to be reminded that "myth," whatever else it might be, is part of "the primary strata of cultural reflection, although their presence may be more or less obvious, more or less influential at various times," as William G. Doty has noted (1980, 532). In addition, these "ideologies in narrative form" that comprise myths are narratives that have lasted. According to Geoffrey Kirk, "they have been passed down from generation to generation" and "to be more than transitory, the kind of tale that is forgotten almost as soon as told, they must possess certain special qualities" (1970, 282). Those "special qualities," as Karen Armstrong delightfully describes them, include "the sense of significance" that members of societies seem "to require," and, through myth are offered "structure and meaning [in] life" (2005, 128).

Myth is, therefore, reiterated, retold, and reworked, constantly within cultures. More to the point, the truth of myth is something that cultures apply to the myths, themselves. William James, one of the founders of American pragmatic philosophy, described "truth" in a philosophical sense that strongly connects to the same notion about the truth of myths. Where the Myth and Symbol scholars, drawing from the functionalism of Malinowski, see myth as essentially a static set of ideologies—which is exactly the critique New Americanists bring to bear in rejecting such scholarship—what James said about "truth" fits far more closely to contemporary understandings of the concept:

> Truth *happens* to an idea. It *becomes* true, is *made* true by events. Its verity *is* in fact an event, a process, the process namely of its verifying itself, its veri*fication*. Its validity is the process of its vali*dation*.
>
> (1997, x; original emphasis)

The contention James puts forward about truth, in other words, is the same contention made here about myth. James's notion of how "truth" becomes "true" is precisely the same in that myth "become true, are made true" within a culture—individual members of a culture make them "true," and in "becoming true" they are made culturally "valid." This pragmatic position sidesteps many of the issues inherent in either Modernist or postmodern conceptions of knowledge as well. As I have previously noted, when this position is put forward "instead of questioning the meaning of 'truth' or [empirical] 'reality,' we see these issues evaporate" and can, instead, concentrate on cultural meaning and experience (Wilhelm and Kapell 2004, 136). What myths do for members of a society, drawing from another work by James on the pragmatic conception of truth, is that they "*become true just in so far as they help us to get into satisfactory relation with other parts of our experience*" (1995, 23; original emphasis). The topics discussed here are experiments in such a process. Americans were offered differing versions of myths and then segments of the population engaged with those myths granting the possibility that they will thus "*become* true." Such a "mythic truth" does not cohere around any conception of a myth's relationship to a measurable and quantifiable reality, because to relate myths to empirical truth in that fashion is to offer judgment of a nonempirical realm of cultural experience by the rules that do not apply. Instead, working through such myths is an attempt for find myths that are, in Rorty's terminology again, "useful to create a better future" (1999, 27).

The empirical critique against the Myth and Symbol school was predicated on the relationship between the products of that school—their books—and some notions of an empirical reality being described. Myths, however, are never predicated on the idea of a relationship to an empirical truth. Indeed, That was never the intent of books such as Charles L. Sanford's *The Quest for Paradise*, Henry Nash Smith's *Virgin Land*, or Leo Marx's *The Machine in the Garden*. Although written in an acceptable style for the period, in the format of a humanistic social scientific work, the Myth and Symbol school's practitioners were far more in tune with the Jamesian notion of "truth" than the empirical social sciences of their day. Not searching for some ultimate capital-*T* "Truth" they were, instead, following a well-trod, long-established path. Like, "French, English, Dutch, Swiss and German spectators, armchair philosophers, curious novelists and intrepid adventurers," before them, they were "ponder[ing] the meaning of America" (Gamble 2012, 171). As Rorty described this viewpoint, "[o]ur responsibility to Truth is not . . . a responsibility to get things right. Rather, it is a responsibility to ourselves to make our beliefs cohere with one another, and to our fellow humans to make them cohere with theirs" (1999, 149). This notion of truth, as contrasted with a form of myth as nontruth, is essentially to judge one-half of the equation—myth—through the guidelines of the other. Where Rorty would argue that such a contrast is possible, saying that "calling a story 'mythical' or 'ideological' would be meaningful only if such stories could be contrasted with an 'objective' story" it is the position of this book that this is a failure to recognize two non-overlapping ways of understanding our world (1998, 11). And myths are an incredibly useful way of such understanding—of creating meaning.

When Mark Schorer described myths as "the instruments by which we continually struggle to make our experience intelligible to ourselves," he was approaching the same idea meant by "truth" in James' pragmatic philosophy. Malinowski's notion that myths provide "social glue" might have been a simplified version of this. That is, it is far too reductionistic in its conclusion. But in contemporary myth studies the notion that "myth can become so diffuse that we must constantly redefine it," as Doty claims, is the same process James meant by "truth *happens* to an idea" (1995, 205). In other words the pragmatic process by which "truth" is established is exactly the same process by which "myths" are made and accepted and the way they are *continually revised* within a culture.

Phenomenologically this could be construed as a significant problem for this work. To note these issues from a slightly different perspective, it is possible to consider how much Myth and Symbol school scholarship was actual mythography—that is, the study of myths—and how much was, in the retelling of these myths, actual mythmaking. However, this is yet another empirical critique. It assumes that there remains an external reality, an objective observer, and a gradual accumulation of data about the myths being studied. American religious studies scholar Milton Scarborough has asked the same question about such studies in general, and his answer is one that could be put forward in reply to all the critiques of the Myth and Symbol school:

> They will admit the presence of myth in their own culture as well as that of others, in the present as well as the past, and increasingly in science as well as religion, but yet regard their own theoretical analyses as independent of myth . . . no theory of myth that does not take into account its own myth-dependency can be fully adequate.
>
> (Scarborough 2002, 52)

Where the American Studies Myth and Symbol scholars presented their conception of myth as essentially functionalist, and as social glue holding together a singular national culture, this work is self-consciously aware that in presenting a different form of mythic awareness it is presenting both a different mythography and is offering a different kind of mythmaking as well.

The main topics of this work, after all, will be presented as a narrative. They will be offered within an ideological frame of reference. They will be advanced both as an analytical exercise and as a retelling of a number of already familiar stories. To accomplish this retelling in a newer framework, then, it becomes necessary to offer a different way of conceptualizing myth. And this new orientation for myth recognizes two essential issues: First, that "myths do not speak consistently in only one voice within a society" (Doty 1980, 553). In other words, all myths are polysemous by their very nature and mean different things—represent different "truths"—to different members and different groups within a society. And, second, but equally important, "myths and their interpretations can be *revolutionary*" (Doty 1995, 197; original emphasis). That is, rather than being conservative structures enforcing social order as Malinowski and the Myth and Symbol scholars would have it, the reworking and retelling of myths within a culture, and between subcultures, however defined, can be a major force for change within a nation such as the United States.

Self-consciously aware of this, however, this work also recognizes that such "mythography cannot remain merely a tool for the objective viewing of things such as myths 'over there' somewhere" (Doty 1995, 219). Thus, the transformational conception of myth which follows is reflexively an attempt to approach notions of American mythologies with the intent on both understanding that there is no single American myth—or even groups of myths—and that those myths that are analyzed here can be seen as part of a significantly larger attempt, on the part of a large and diverse society, to both understand itself and remake that understanding again, and again, through a continual reworking of the mythic formulae available.

A TRANSFORMATIONAL CONCEPTION OF MYTH

American Studies scholars may have used *myth* in a variety of ways, and for a similar variety of purposes, but all agreed with Malinowski, Cassirer, Eliade, and others that myth was, essentially, conservative. That is, myth created "social glue" or offered cultural continuity by essentially limiting individual options. On the very last page of the epilogue of *The Machine In The Garden* Leo Marx speaks of that conservative "social glue" paradigm of myth used in American Studies. Explaining at once both his reasoning for writing his book as he did and for what his topic meant for the present—very much the exact criticism of "presentism" that would eventually be applied to the work—Marx noted the "inability of our writers to create a surrogate for the ideal of the middle landscape," as a central point of his analysis. The failure of writers such as Melville, Twain, and Fitzgerald to discover a synthesis between "the machine" and "the garden" in their works, however, "can hardly be accounted artistic failure" (Marx 1964/2000, 364–365). In searching for that synthesis in the works of classic American literature Marx is left with but one conclusion:

> By incorporating in their work the root conflict of our culture, they have clarified our situation. They have served us well. To change the situation we require new symbols of possibility, and although the creation of those symbols is in some measure the responsibility of artists, it is in greater measure the responsibility of society . . . [it] belongs not to art but to politics.
>
> (Marx 1964/2000, 365)

This call for "new symbols of possibility" is a call to find new myths, with new meanings as well. The statement foreshadows the rejection of the Myth and Symbol school that was just a few short years away, and it also issues a call for a new mythic awareness that, ironically, no field—least of all American Studies—would pick up. It is also very much an extended misreading of the way myths work, change, and alter the social landscape. It was not "new symbols of possibility" that would engulf the American society in the remainder of the 1960s; it was an attempt to reimagine what the preexisting symbols could now be used for. Calling for "new symbols of possibility" just as the United States military was about to expand its presence in Vietnam, just as NASA's Gemini missions began two years of extensive preparatory launches that would lead directly to Apollo and the lunar landing remains a failure not to notice the symbols of possibility that were already present and already being haggled over within American society.

Since 1964, when Marx's words were published as the last statement of the last canonical work of the Myth and Symbol school, American Studies and related disciplines have fully rejected "myth" as a positive analytical tool with the advent of the New Left and New Social history. Postmodern myth scholars, however, have not been dissuaded from continuing to use "myth" for examinations of narratives of transition and transformation. That is, they have continued to use "myth" for understanding cultural change and how cultural change shapes new interpretations of existing myths. Nor have the arguments in favor of a reimagining of what myth is, and how to use it, been limited to specific disciplines. Indeed, the call for the use—or at least acknowledgment of—myth comprises minority views in a number of fields. In his 1985 presidential address to the American Historical Association William H. McNeill called for an approach quite similar that the one advocated here. "All truths are general," he noted. But a realization of the use of such an understanding of "truths" being inherent in the cultural myths of people would allow "a more rigorous and reflective epistemology, [with which] we might also attain a better historiographical balance between Truth, truths, and myth" (McNeill 1986b, 8). McNeill, a well-known global historian, knew that his approach would not be well liked by colleagues. Calling it "mythistory," and proclaiming, "I do not expect the term to catch on in professional circles," McNeill tried to connect the nonempirical notion of "myth" with the traditionally empirical conceit of professional historians. He was right, however. The term—and the idea it represented—would not catch on in professional circles. This work, however, does accept his central premise, that

> historians' truths, like those of scientists, evolve across the generations, so that versions of the past acceptable today are superior in scope, range, and accuracy to versions available in earlier times. But such evolution is slow, and observable only on an extended time scale, owing to the self-validating character of myth.
>
> (McNeill 1986b, 9)

For McNeill, what professional historians do is largely take part in validating societal myths. This is, of course, something empirically minded historians would find anathema. But, in taking part of such validations, historians create the "ideology in narrative form" that is central to the mythical understandings of a people. As a result, when those mythic understandings are in the process of radical change, and academics fail to notice the mythical underpinnings of those changes, the purpose of "history" becomes, at best, suspect. Although McNeill's hope for a kind of "mythistory" to overcome this problem fell largely on deaf ears amongst the crowd of the American Historical Association, other fields have continued to call for an understanding in a similar vein.

John Girling, a political sociologist, has offered a rough sketch of the relationship between self-validating character of myths and social change: "The inspirational character of myths is evident in turbulent times. It is this that enabled believers to adapt to, and even to shape, the 'reality' of powerful, impersonal forces at work in society" (1993, 170). For Girling this is not a question of Marx's "new symbols of possibility" so much as it is a question of reworking the meanings of already existing symbols. As a result, traditional mythic understandings can be reworked to provide new ways of viewing the world and to alter the worldviews of people in society. Myths, in other words, need not be wholly rejected and replaced with "new symbols" for those myths to offer transformational capabilities.

Presenting myths as transformational rather than static, as capable of promoting social changes rather than conservatively hindering them is precisely how this work is organized. William G. Doty notes that "myths may actualize parts of the underlying cultural worldview," which is exactly what Myth and Symbol scholars in American Studies hoped to explain (1980, 534). This project goes further, agreeing with Doty that

> [s]eldom does a single myth actualize the entire worldview, since that seems to require a collection of many interlocked stories, a canon rather than one sample. In the processes of transmission constant change and adaption to new or changed context seem to be normal. A particular myth may undergo addition and expansion, deletion or substitution, of mythemic units . . . [And] transformations and variations may occur within a network.
>
> (1980, 534)

In detailing experiments in mythmaking, then, this work offers an examination of Doty's idea that myths can but "anticipate *forms of the future* as they determine and shape ideals and goals for both individual and society" (2000, 26; original emphasis). In doing so, society is in the process of testing these mythic forms to see if, as James would put it, "truth *happens* to them."

Furthermore, the four major topics under consideration in this work do not represent the whole of some Platonically idealized national worldview of the entire population of the United States. But, they do all represent aspects of similar mythical understandings, that can be shown to speak across the events of "constant change" in the late 1960s and 1970s, each undergoing "addition and expansion" as well as "deletion or substitution." This remains the single greatest reason to offer them, together, for analysis. Such an analysis only works, however, if the previous myth scholarship that dominated the study of "myths and symbols" in the period is significantly adjusted to allow the discussion of "transformations and variations."

Stephen H. Daniel described this method as showing that "mythic expression . . . opens the possibility that the relational or ontological character of the world might be changed" (1990, 6). As various cultural events and products—including those analyzed here—affected the multiple cultures of the United States, and as individuals, groups, and the polity itself reacted, the myths become, again in Doty's phrase, "important change-agents, as communities move into different socio-political realities" (2002, 52). The fact that the topics here culturally engaged through the events surrounding Vietnam and the moon landing, the steadily increasing popularity of *Star Trek* and the initial popularity of Gerard K. O'Neill's High Frontier—that is, these things as historical and cultural occurrences, and things to which American culture reacted to—allows for a method of myth analysis which posits myth as a way in which cultures work to change themselves. Thus, the topics here are considered as part of the transformational mythscape beginning at the very end of the 1960s that *must* be examined *together* if a more complete understanding of the period is ever to be achieved. This is true even if these mythic expressions do not constitute a kind of singular weltanschauung of the United States but offer merely insight into the cultural play around frontier myths occurring for some part of the American polity.

Israeli historian Joseph Mali describes the concept that propels this project in his book *Mythistory*. If critiques of previous scholarship in the Myth and Symbol school are based on anything at all, then it is based on the notion, described by Mali generally, that "history was a serious and reliable explanation of what actually happened insofar as empirical sources and discourses were veritable." For Mali, as well as for this study, "the critical task of this historiography, or *mythistory*, is to reappraise these stories as inevitable, and ultimately valuable, histories of personal and communal identity" (2003, xii; original emphasis).

Put from only a slightly difference perspective, then, the central theme of this project is that without an examination of the transformational possibilities inherent in the various myths of the period an understanding of the period, itself, will always be incomplete. Again, John Girling describes this perspective in a way that offers insight into all of the topics under consideration here:

> Myths are symbolic representations of critical changes; they do not "explain" these changes. What an interpretation of myths can do is to help to understand why people (ourselves) respond in certain ways to the situations they encounter, that is, why they behave with such

> intensity, commitment, and perseverance; either creatively mobilizing popular energies, even against rational odds . . . or else mobilizing the same psychic energies, but on a destructive course.
>
> (1993, 170)

This work examines and emergent mythscape beginning in the late 1960s, where different versions of cultural myths were also talking at such cross-purposes. But where scholarship in such a situation is held back, it is such cultural discussions among various forms of myths that can actually begin to propel a culture forward. That is, as Laurence Coupe noted in a general context, while myths such as those discussed here may reference an "age that knows more about chaos than it does about cosmos," that does not mean such myths are only about the past. "If myth is ultimately about possibility," Coupe continues, then beginning in the past but pointing to the future "will always be appropriate" (2009, 89).

The American experience of Vietnam seemed, at the time, to preclude such mythic experiments and fully demand, as Marx put it, wholly "new symbols of possibility." But at the same time, American culture was experimenting with other forms of traditional myths in NASA, *Star Trek* and Gerard K. O'Neill's High Frontier. These experiments were not merely static myths of unchanging conservative ideals but were actually Robertson's "imperatives for present and future" and Doty's "forms of the future." As myths such as these developed within American culture the culture, itself, responded by working with them in an attempt to ascertain if they could offer William James's conception of truth. The question is not, "Are these reasonable representations of an empirical, objective reality?"

The questions are, Would "truth *happen* to" any of these myths? Would they "*become* true?"

NOTES

1 The first few paragraphs in this chapter ironically reference Henry Nash Smith's famous opening statement about "myth and symbol" in *Virgin Land* (1971, see especially xi–xii, 4).

2 This quote is from a prepared lecture published after Malinowski's death and was initially labeled simply "lecture one." Its exact date therefore remains uncertain.

3 Originally from a review of *The Mythology of All Races* (Uno Holmberg) this was originally published in 1928.

4 The quote here is drawn from page 2 but a more general description can be found in pages 105–119.

5 Limerick is citing her own document, a one-page summation prepared for the 1989 conference.

6 Fitzpatrick de-emphasizes the notion of a single, unitary school of "consensus historians" quite strongly; however she does capture an excellent analysis of why the term came to be strongly identified with the period.

7 Decker's argument is significantly more complex than is implied here and hinges on Kuklick's definition of *empiricism*.

8 This discussion of myth is adapted freely from my own previous, but much briefer, analyses. See Kapell (2010b, 3–5); M. W. Kapell (2006; 2010a, 2015, 223–229), and M. W. Kapell and Elliott (2013).
9 Rowe (2002, 51–63) offers only a brief mention of pragmatism, itself, but offers significant insight on the problems I am attempting to avoid here.

WORKS CITED

Armstrong, Karen. 2000. *The Battle for God*. New York: Alfred A. Knopf.
———. 2005. *A Short History of Myth*. Edinburgh: Canongate.
Bercovitch, Sacvan. 1986a. "America as Canon and Context: Literary History in a Time of Dissensus." *American Literature* 58: 9.
———. 1986b. "The Problem of Ideology in American Literary History." *Critical Inquiry* 12 (4): 631–53.
———. 1991. "Investigations of an Americanist." *The Journal of American History* 78 (3): 972–87.
Billington, Ray Allen. 1974. *Westward Expansion: A History of the American Frontier*. 4th ed. New York: Macmi.
Bogue, Allan G. 1994. "Frederick Jackson Turner Reconsidered." *The History Teacher* 27 (2): 195–221.
Brooks, Van Wyck. 1918. "On Creating a Usable Past." *The Dial* LXIV (7): 337–41.
Carpenter, Ronald H. 1983. *The Eloquence of Frederick Jackson Turner*. San Marino: Huntington Library.
Cochran, Thomas C. 1972. *Social Change in America: The Twentieth Century*. New York: Harper & Row.
Cohen, Percy S. 1969. "Theories of Myth." *Man*, New Series, 4 (3): 337–53.
Coupe, Laurence. 2009. *Myth*. New York: Routledge.
Cronon, William. 1987. "Revisiting the Vanishing Frontier: The Legacy of Frederick Jackson Turner." *The Western Historical Quarterly* 18 (2): 157–76.
———. 1991. "Turner's First Stand: The Significance of Significance in American History." In *Writing Western History: Essays on Major Western Historians*, edited by Richard W. Etulain, 73–101. Albuquerque: University of New Mexico Press.
Csapo, Eric. 2005. *Theories of Mythology*. Malden, MA: Blackwell Pub.
Daniel, Stephen H. 1990. *Myth and Modern Philosophy*. Philadelphia: Temple University Press.
Davies, Stephen. 2003. *Empiricism and History*. Houndsmills, Basingstoke, Hampshire: Palgrave.
Decker, Jeffrey Louis. 1992. "Dis-Assembling the Machine in the Garden: Antihumanism and the Critique of American Studies." *New Literary History* 23 (2): 281–306.
Doty, William G. 1980. "Mythophiles' Dyscrasia: A Comprehensive Definition of Myth." *Journal of the American Academy of Religion* XLVIII (4): 531–62.
———. 1995. "Silent Myths Singing in the Blood: The Sites of Production and Consumption of Myths in a 'Mythless' Society." In *Picturing Cultural Values in Postmodern America*, edited by William G. Doty, 187–220. Tuscaloosa: University of Alabama Press.
———. 2000. *Mythography: The Study of Myths and Rituals*. 2nd ed. Tuscaloosa: University of Alabama Press.

———. 2001. "Modern and Postmodern Mythic Existence." Lecture presented at the Myth Symposium, Indiana University Folklore Institute, March 30.

———. 2002. "Myth and Postmodernist Philosophy." In *Thinking Through Myths: Philosophical Perspectives*, edited by Kevin Schilbrack, 142–57. New York: Routledge.

———. 2004. *Myth: A Handbook*. Greenwood Folklore Handbooks. Westport, CT: Greenwood Press.

Etulain, Richard W. 1999. "The Frontier and American Exceptionalism." In *Does the Frontier Experience Make America Exceptional?*, edited by Richard W. Etulain, 3–14. New York: Bedford/St. Martin's.

Fabian, Ann. 1996. "Review: Back to 'Virgin Land.'" *Reviews in American History* 24 (3): 542–53.

Fischer, David Hackett. 1970. *Historians' Fallacies: Toward a Logic of Historical Thought*. New York: Harper & Row.

Fisher, Marvin. 1962. "Review: The Quest for Paradise." *American Quarterly* 14 (2): 211–12.

Fisher, Philip. 1991a. "Introduction: The New American Studies." In *The New American Studies: Essays from Representations*, edited by Philip fisher, vii–xxii. Berkeley: University of California Press.

———, ed. 1991b. *The New American Studies: Essays from Representations*. Representations Books 5. Berkeley: University of California Press.

Fitzpatrick, Ellen F. 2002. *History's Memory: Writing America's Past, 1880–1980*. Cambridge: Harvard University Press.

Gamble, Richard M. 2012. *In Search of the City on a Hill: The Making and Unmaking of an American Myth*. New York: Continuum. http://www.amazon.com/Search-City-Hill-Unmaking-American/dp/1441162321/ref=sr_1_1?ie=UTF8&qid=1442618200&sr=8–1&keywords=gamble+city+on+a+hill.

Girling, J. L. S. 1993. *Myths and Politics in Western Societies: Evaluating the Crisis of Modernity in the United States, Germany, and Great Britain*. New Brunswick, NJ: Transaction Publishers.

Hatab, Lawrence J. 1990. *Myth and Philosophy: A Contest of Truths*. La Salle, IL: Open Court.

Heyne, Eric. 1992. "Lasting Frontier: Reinventing America." In *Desert, Garden, Margin, Range: Literature on the American Frontier*, edited by Eric Heyne, 3–15. New York: Twayne Publishers.

Higham, John. 1971. "The Cult of the 'American Consensus': Homogenizing Our History." In *The Shaping of Twentieth-Century America*, edited by Richard M. Abrams and Lawrence W. Levine, 699–709. Boston: Little, Brown.

———. 1989. *History: Professional Scholarship in America*. Updated pbk. ed. Baltimore: Johns Hopkins University Press.

Hofstadter, Richard. 1979. *The Progressive Historians: Turner, Beard, Parrington*. Chicago: University of Chicago Press.

Horwitz, Richard P. 2001. "America Studies: Approaches and Concepts." In *Encyclopedia of American Studies*, edited by George Kurian, Miles Orvell, Johnnella Butler, and Jay Mechling, 112–18. Bethel, CT: Grolier Publishing Company. http://myweb.uiowa.edu/rhorwitz/asconcepts.htm.

Hughes, Richard T. 2003. *Myths America Lives By*. Urbana: University of Illinois Press.

Jacobs, Wilbur R. 1965. *Turner, Bolton, and Webb: Three Historians of the American Frontier*. Seattle: University of Washington Press.

James, William. 1960. *The Will to Believe, and Other Essays in Popular Philosophy, and Human Immortality*. New York: Dover Publications.

———. 1995. *Pragmatism*. Dover Thrift Editions. New York: Dover Publications.

———. 1997. *The Meaning of Truth*. Amherst, NY: Prometheus Books.

Jewett, Robert, and John Shelton Lawrence. 1977. *The American Monomyth*. Garden City: Anchor Press.

Jones, Karen R., and John Wills. 2009. *American West: Competing Visions*. Edinburgh: Edinburgh University Press.

Kapell, Matthew. 2002. "Civilization and Its Discontents: American Monomythic Structure as Historical Simulacrum." *Popular Culture Review* 13 (2): 129–36.

Kapell, Matthew, and Suzanne Becker. 2005. "Patriarchy, the Christian Romance Novel and the 'Ecosystem of Sex.' " *Popular Culture Review* 16 (1): 147–55.

Kapell, Matthew Wilhelm. 2006. "Conclusion: Finding Myth in the History of Your Own Time." In *Finding the Force of the Star Wars Franchise: Fans, Merchandise and Critics*, edited by Matthew Wilhelm Kapell and John Shelton Lawrence, 269–76. New York: P. Lang.

———. 2010a. "Conclusion: The Hero with a Thousand Red Shirts." In *Star Trek as Myth: Essays on Symbol and Archetype at the Final Frontier*, edited by Matthew Wilhelm Kapell, 213–19. Jefferson, NC and London: McFarland & Co.

———. 2010b. "Introduction: The Significance of the Star Trek Mythos." In *Star Trek as Myth: Essays on Symbol and Archetype at the Final Frontier*, edited by Matthew Wilhelm Kapell, 1–16. Jefferson, NC and London: McFarland & Co.

———. 2015. "Conclusion: Adaptation or Translation?" In *The Fantastic Made Visible: Essays on the Adaptation of Science Fiction and Fantasy from Page to Screen*, edited by Kapell, Matthew Wilhelm and Ace G. Pilkington, 223–29. Jefferson, NC and London: McFarland & Co.

Kapell, Matthew Wilhelm, and Andrew B. R. Elliott. 2013. "Conclusion(s): Playing at True Myths, Engaging with Authentic Histories." In *Playing with the Past: Digital Games and the Simulation of History*, edited by Matthew Wilhelm Kapell and Andrew B. R. Elliott, 357–69. New York: Bloomsbury Academic.

Kirk, G. S. 1970. *Myth: Its Meaning and Functions in Ancient and Other Cultures*. v. 40. Cambridge: Cambridge University Press.

Kolodny, Annette. 1984. *The Lay of the Land: Metaphor as Experience and History in American Life and Letters*. Chapel Hill: The University of North Carolina Press.

Kuklick, Bruce. 1972. "Myth and Symbol in American Studies." *American Quarterly* 24 (4): 435–50.

Leach, E.R. 1957. "The Epistemological Background to Malinowski's Empiricism." In *Man and Culture: An Evaluation of the Work of Bronislaw Malinowski*, edited by Raymond Firth, 119–38. London: Routledge & K. Paul.

Lewis, R. W. B. 1955. *The American Adam: Innocence, Tragedy, and Tradition in the Nineteenth Century*. Chicago: University of Chicago Press.

Limerick, Patricia Nelson. 1991. "What on Earth Is the New Western History?" In *Trails: Toward a New Western History*, edited by Limerick, Patricia Nelson, Clyde A. Milner, and Rankin, Charles E., 81–8. Lawrence: University of Kansas Press.

———. 1995. "Turnerians All: The Dream of a Helpful History in an Intelligible World." *American Historical Review* 100 (3): 697–716.

Limerick, Patricia Nelson, Clyde A. Milner, and Charles E. Rankin. 1991. "Preface." In *Trails: Toward a New Western History*, ix–xv. Lawrence: University of Kansas Press.

Lincoln, Bruce. 1999. *Theorizing Myth: Narrative, Ideology, and Scholarship*. Chicago: University of Chicago Press.

Madsen, Deborah L. 1998. *American Exceptionalism*. BAAS Paperbacks. Edinburgh: Edinburgh University Press.

Mali, Joseph. 2003. *Mythistory: The Making of a Modern Historiography*. Chicago: University of Chicago Press.

Malinowski, Bronislaw. 1923. "The Problem of Meaning in Primitive Languages." In *The Meaning of Meaning*, edited by C. K. Ogden, 296–336. New York: Hartcourt, Brace.

———. 1954. "Myth in Primitive Psychology." In *Magic, Science and Religion and Other Essays*, 93–148. New York: Doubleday.

———. 1963a. "Myth as a Dramatic Development of Dogma (essays)." In *Sex, Culture and Myth*, 245–55. London: Rupert Hart-Davis.

———. 1963b. "The Life of Myth." In *Sex, Culture and Myth*, 289–94. London: Rupert Hart-Davis.

Marx, Leo. 1964. *The Machine in the Garden: Technology and the Pastoral Ideal in America*. New York: Oxford University Press.

———. 2005. "On Recovering the 'Ur' Theory of American Studies." *American Literary History* 17 (1): 118–34.

McConnell, Frank D. 1979. *Storytelling and Mythmaking: Images from Film and Literature*. New York: Oxford University Press.

McNeill, William H. 1986a. "Mythistory, or Truth, Myth, History, and Historians." In *Mythistory and Other Essays*, 3–22. Chicago: University of Chicago Press.

———. 1986b. "Mythistory, or Truth, Myth, History, and Historians." *The American Historical Review* 91 (1): 1–10.

McVeigh, Stephen. 2007. *The American Western*. Edinburgh: Edinburgh University Press.

Mogen, David. 1993. *Wilderness Visions: The Western Theme in Science Fiction Literature*. 2nd ed., revised and expanded. Rockville, MD: Wildside Press.

Noble, David W. 1965. *Historians Against History: The Frontier Thesis and the National Covenant in American Historical Writing Since 1830*. Minneapolis: University of Minnesota Press.

———. 1968. *The Eternal Adam and the New World Garden: The Central Myth in the American Novel Since 1830*. New York: Braziller.

———. 2002. *Death of a Nation: American Culture and the End of Exceptionalism*. Critical American Studies Series. Minneapolis: University of Minnesota Press.

Paul, Heike. 2014. *The Myths That Made America: An Introduction to American Studies*. Bielefeld, Nordrhein-Westfalen: Transcript-Verlag.

Paxson, Frederic L. 1967. *History of the American Frontier, 1763–1893*. Dunwoody, Georgia: N.S. Berg.

Pease, Donald E., and Robyn Wiegman. 2002a. "Futures." In *The Futures of American Studies*, edited by Donald E. Pease and Wiegman, Robyn, 1–42. Durham: Duke University Press.

———., eds. 2002b. *The Futures of American Studies*. New Americanists. Durham: Duke University Press.

Rorty, Richard. 1963. "Empiricism, Extensionalism, and Reductionism." *Mind*, New Series, 72 (286): 176–86.

———. 1998. *Achieving Our Country: Leftist Thought in Twentieth-Century America*. The William E. Massey, Sr. Lectures in the History of American Civilization 1997. Cambridge, MA: Harvard University Press.

———. 1999. *Philosophy and Social Hope*. New York: Penguin Books.

Rowe, John Carlos. 2002. *The New American Studies*. Minneapolis: University of Minnesota Press.

Sanford, Charles L. 1952. "The Garden of America." *Modern Review* XCII: 23–32.

———. 1958. "The Intellectual Origins and New-Worldliness of American Industry." *The Journal of Economic History* 18 (1): 1–16.

———. 1961. *The Quest for Paradise: Europe and the American Moral Imagination*. Urbana: University of Illinois Press.

Scarborough, Milton. 2002. "Myth and Phenomenology." In *Thinking Through Myths: Philosophical Perspectives*, edited by Kevin Schilbrack, 46–64. New York: Routledge.

Schorer, Mark. 1968. "Appendix: The Necessity of Myth." In *Myth and Mythmaking*, edited by Henry Alexander Murray, 354–58. Boston: Beacon Press.

Singly, Carole J. 1992. "Catherine Maria Sedgwick's Hope Leslie: Radical Frontier Romance." In *Desert, Garden, Margin, Range: Literature on the American Frontier*, edited by Eric Heyne, 110–22. New York: Twayne.

Sklar, Robert. 1970. "American Studies and the Realities of America." *American Quarterly* 22 (2): 597–605.

Slotkin, Richard. 1973. *Regeneration Through Violence: The Mythology of the American Frontier, 1600–1860*. Norman: University of Oklahoma Press.

———. 1981. "Nostalgia and Progress: Theodore Roosevelt's Myth of the Frontier." *American Quarterly* 33 (5): 608–37.

———. 1993. *Gunfighter Nation: The Myth of the Frontier in Twentieth-Century America*. New York: HarperPerennial.

———. 1998. *The Fatal Environment: The Myth of the Frontier in the Age of Industrialization, 1800–1890*. Norman: University of Oklahoma Press.

Smith, Henry Nash. 1957. "Can 'American Studies' Develop a Method?" *American Quarterly* 9 (2): 197–208.

———. 1971. *Virgin Land: The American West as Symbol and Myth*. Cambridge, MA: Harvard University Press.

Spiller, Robert E. 1960. "American Studies, Past, Present, and Future." In *Studies in American Culture: Dominant Ideas and Images*, edited by Joseph J Kwiat and Mary C Turpie, 207–20. Minneapolis: University of Minnesota Press.

Strenski, Ivan. 1987. *Four Theories of Myth in Twentieth-Century History: Cassirer, Eliade, Lévi-Strauss, and Malinowski*. 1st ed. Iowa City: University of Iowa Press.

Tate, Cecil F. 1973. *The Search for a Method in American Studies*. Minneapolis: University of Minnesota Press.

Thompson, Gerald. 1991. "Another Look at Frontier/Western Historiography." In *Trails: Toward a New Western History*, edited by Patricia Nelson Limerick, Clyde A. Milner, and Rankin, Charles E., 89–96. Lawrence: University of Kansas Press.

Trachtenberg, Alan. 1979. *Brooklyn Bridge: Fact and Symbol*. Phoenix ed. Chicago: University of Chicago Press.

———. 1982. *The Incorporation of America: Culture and Society in the Gilded Age*. New York: Hill and Wang.

———. 1984. "Myth and Symbol." *Massachusettes Review* 25: 667–73.

Turner, Frederick Jackson. 1903. "The Significance of the Louisiana Purchase." *The American Monthly Review of Reviews* 27 (1): 584.

———. 1917. "The Significance of the North Central States in the Middle of the Nineteenth Century." Presented at the American Historical Association, Philadelphia.

———. 1925. "The Significance of the Section in American History." *Wisconsin Magazine of History* 8 (3): 255–80.

———. 1961. "The Significance of History." In *Frontier and Section*, edited by Ray Allen Billington, 2–27. Englewood Cliffs, NJ: Prentice Hall.

———. 1999. "The Significance of the Frontier in American History." In *Does the Frontier Experience Make America Exceptional?*, edited by Richard W. Etulain, 17–43. Boston: Bedford/St. Martin's.

Ward, John William. 1955. *Andrew Jackson: Symbol for an Age*. New York: Oxford University Press.

———. 1962. "Review: The Quest for Paradise." *The Mississippi Valley Historical Review* 48 (4): 723–25.

Webb, Walter Prescott. 1931. *The Great Plains*. Boston: Ginn and company.

———. 1952. *The Great Frontier*. Boston: Houghton Mifflin.

Welbourn, Fred. 1964. "Myth, Rationalism and Empiricism." *Transition*, no. 16 (January): 40–42.

Wilhelm, Stephanie J., and Matthew Kapell. 2004. "Visions of Hope, Freedom of Choice, and the Alleviation of Social Misery: A Pragmatic Reading of the Matrix." In *Jacking in to the Matrix Franchise: Cultural Reception and Interpretation*, edited by Matthew Kapell and William G. Doty, 125–40. New York: Continuum.

Wise, Gene. 1973. *American Historical Explanations: A Strategy for Grounded Inquiry*. Homewood, IL: Dorsey Press.

———. 1979. "'Paradigm Dramas' in American Studies: A Cultural and Institutional History of the Movement." *American Quarterly* 31 (3): 293–337.

Wrobel, David M. 1993. *The End of American Exceptionalism: Frontier Anxiety from the Old West to the New Deal*. Lawrence: University Press of Kansas.

3 Vietnam, *The Forever War*, and the Shattering of American Myth

Where is John Wayne now that we really need him?

—Joe Haldeman, *The Forever War*, 1974 (2004c, 179)

Joe Haldeman did not want to go to Vietnam.

At twenty-four he had avoided the draft by being a full-time student. But, on graduation, that route was no longer available. And in 1967, he had done just about everything else a draft-eligible American could do. He had applied for a civilian job with the navy that would have exempted him from military service, and as he put it in an autobiographical sketch, he originally wrote for the Gale *Contemporary Authors* series:

> I had applied to the Peace Corps as well, saying I'd rather spend six years digging ditches in Africa than one year killing people in Vietnam, but they never responded. Earlier, I had petitioned for conscientious objector status, but the draft board wouldn't even start the process until you had a letter from your minister, and atheists don't have ministers. Stupidly, I let it go at that. (2004b)

In the end, returning from a family vacation to Canada to find his summons, Haldeman finally decided to accept his draftee status for a relatively simple reason:

> My formal rationale for not going to jail or Sweden was that I was still determined to become a spaceman, which for the foreseeable future would be a government job. In the late sixties, NASA had a "Scientist as Astronaut" program: they would take a Ph.D. in science or engineering and teach him (or her, in theory) how to fly jets, which was easier than getting a jet jockey to go back to graduate school. (2004b)

NASA's "Scientist as Astronaut" program, and the desire to become a "spaceman," would encourage Haldeman to turn away from his pacifism, his anti–Vietnam War sentiment and, like another scientifically trained person who is discussed in this work—Gerard K. O'Neill—hold out hope for a future in space. But even in the army he would work hard to avoid direct combat because, as he put it, "my pacifist sympathies are pretty obvious, at least to my friends" (Haldeman, 1993, 135). He "waved [his] physics degree at the Army and offered to sign up for four years of nuclear power plant operation," but that, too, failed. The Army deemed him "overqualified" (ibid., 136). Finally, with no other options seemingly available, the army sent him for his year in Vietnam on February 29, 1968, just barely at the end of the Tet Offensive, which had been launched on January 30 and composed a major turning point in the war. With a Bachelor of Science in astronomy and physics from the University of Maryland, Joe Haldeman became a combat engineer in Vietnam—and not just a combat engineer but a "Combat Engineer, Pioneer." Haldeman described his initiation to the role: "they told us [it] meant we were sort of like infantry, but too dumb to carry a rifle, so they gave us a shovel" (2004b). He did not last the full year but made it until September when an officer ordered Haldeman and three other combat engineers to guard a captured enemy "DX pile"—a collection of enemy explosives in need destruction. In his poem "DX" Haldeman recalled of this duty, "We'd been lucky / No shooting. / Just a pile of / explosive / leftovers / to dispose of" (2005c, 166–167). But he would describe the outcome of that duty:

> It was boobytrapped, with a time-delay or a radio-frequency fuse. After we'd been standing guard for thirty or forty minutes, it suddenly blew. If it had worked well, it might have killed dozens of people, maybe a hundred . . . But only a small fraction of the ordnance actually cooked, so the only people killed were engineers: all of my squad but me. (2004b)

That instant—and the rest of his time "in country"—would become a source of material for a professional writing career that would span Haldeman's return from Vietnam until the present, include multiple award-winning novels and short stories and eventually a writing professorship at the Massachusetts Institute of Technology. It was, as soldiers called it, a "million-dollar wound"—that is, in Haldeman's phrasing, a wound to a soldier "that would be bad enough to take him out of combat, but not serious enough to ruin his life." (2004b) He would describe his "million-dollar wound" as an event in which "twenty or thirty pounds of Chinese high explosive goes off a couple of yards away [and] you don't hear a thing. [You] find yourself suddenly horizontal" (1979, 3). He detailed it this way:

> In one instant I sustained a couple of hundred small puncture wounds, a couple of dozen serious bullet and shrapnel wounds, and one killer—a .51 caliber machine gun bullet to the thigh. The surgeon later told me he'd never taken one out of a person before. They usually just blow off your leg and keep going. (2004b)

The rest of Haldeman's Purple Heart–winning experience, as he wrote of it in the second person, was both unique and far too common:

> You find the green jungle suddenly grey with smoke and crimson with sprayed blood; try to stand up and realize you can't feel your legs; look down to see a femoral artery pumping; try to staunch it with both hands; can't spare a hand to wipe the blood out of your eye; try wiping it on your shoulder, which doesn't help, it's like raw hamburger; holler for a medic and see that your medic is lying beside you with both legs blown off, good friend, his dying screams a delicate threnody in a dull, silent universe.
>
> (1979, 3)

In the end, then, Joe Haldeman went to Vietnam, got his medals, and returned home, and each passing year has made the experience of Vietnam for the United States, "more obscure, which is not a bad thing" (2011a). It would be a gross oversimplification to suggest that, in Joe Haldeman's personal experiences as a draftee and a combat engineer in Vietnam, there could be found a general narrative of the "every soldiers'" experiences. For each American—and each Vietnamese citizen as well—had her or his own experiences, some common and others unique, during the war.

Like many American veterans of Vietnam, Haldeman did turn to writing about his experiences and the experiences of his comrades there. Unlike most, however, Haldeman's writing was done as fantastic literature—as science fiction. As science fiction his *The Forever War*, which was also his thesis for his Master of Fine Arts from the Iowa Writer's Program at the University of Iowa, remains both one of the most important works about Vietnam and, outside of science fiction critics, one of the least examined.

By his own admission Haldeman wrote as much as could be expected while in Vietnam: "roughly half in combat and half in hospitals, but it was almost all in the form of letters" (2005e, 2). Eventually, some of that writing would become the short novel *War Year*, which he called a "realistic, naturalistic novel simply about Vietnam" (1993, 146). His later novel *1968*, which offers both the realism of *War Year* and an insightful examination of the American home front, continued his exploration of the effect of the Vietnam War (2005a). Or, as Haldeman described *1968*, it was "an exploration of a year I felt I had missed, which became a pivotal year in American culture" (2005d, 171). Like Haldeman the protagonist of *1968* is the only survivor of his unit, and like Haldeman the protagonist of *War Year* spends much of his noncombat time writing letters home. Neither work is strictly autobiographical, as Haldeman has pointed out repeatedly; the characters are all composites of people the author knew during or after the war. But realism or naturalism would not suffice to contain the messages about Vietnam Haldeman wanted to write about. For that he would need what H. Bruce Franklin, the American literature and cultural scholar, has called "the principal non-realistic imaginative mode of our historical epoch" (1995, 4). And as an imaginative mode, Vietnam would present a useful starting point for fiction. Writing an introduction for their collection of science fiction and fantasy fiction based on the Vietnam War, Jack Dann and Jeanne Van Buren Dann argued that the "Vietnam War was so psychologically cyclonic and horrific in its effect on individuals that perhaps the dark personal 'truth' of their experience can best be reflected through the devices of metaphor and fantasy" (1987, 14). The Dann's collection—which included Haldeman's poem "DX"—offered an inkling of that "truth" through its title, *In the Field of Fire*. They hoped to offer "another level of meaning and truth to the war," and the collection thus avoids realism in favor of a different "imaginative mode" (1987, 14). Haldeman's *Forever War* would be an earlier, significant effort on the same path. For Haldeman, then, his "truth" of Vietnam would need to be expressed as science fiction.

By turning to science fiction Haldeman would offer both his most widely known exploration on Vietnam, and for our purposes here, his most mythical. Canadian media scholar Tim Blackmore, who interviewed Haldeman in the early 1990s about his Vietnam writing, has claimed that "Haldeman's reasons for his presence in the world of science fiction are clarified by his feeling that traditional fiction is too limited to hold the mythic story" (1993, 133). And in the creation of such mythic stories Haldeman actively used the traditional narrative structures of science fiction but bends them to his own purposes. Using many—if not most—of the conventions of science fiction, Haldeman's 1974 novel *The Forever War* fundamentally reworks those long-established tropes for his own purposes and, as a result, fully undermines them for narrative effect. And if American science fiction is, as David Mogen has claimed, a "contemporary reworking of traditional themes of frontier literature," then Haldeman's science fictional view of Vietnam found none of those traditional mythic hopes on that frontier (1993, 14). There was no wealth, nor opportunity, nor paradise on Haldeman's frontier of space. Although Haldeman uses all of the themes, tropes, and clichés of American science fiction in its frontier form, he continually undermines all of the narrative traditions he employs. And since American science fiction was born, as the American scholar Carl Abbott has suggested, "on the ideas, tropes, and narratives that are rooted in understandings of a western regional past and present," then Haldeman's uses of those "rooted understandings" inverts that tradition in precisely the same way that the American experience in Vietnam also seemed to invert frontier ideology (2006, 18). Where Abbott can find in American science fiction, then, a "seamless connection of centuries, a set of stories that links the nineteenth and twentieth centuries to the twenty-first century and beyond," Haldeman's *Forever War* offers the illusion of those same connections by utilizing the traditional tropes of the genre, but in reality actually offers them, in part, for the purpose of disruption of genre traditions (2006, 18). This is because instead of finding the usual versions of what another American historian turned science fiction critic, William H. Katerberg calls the "essentially American" aspect of the genre with "its optimistic emphasis on new frontiers," Haldeman finds little that is grandly optimistic in his future frontier (2008, 205). What Haldeman offers in *The Forever War*, instead, is violence. And, most important perhaps, on the science fiction frontier of his award-winning novel, he found alienation.

If Joe Haldeman's experiences in Vietnam cannot be considered a general statement on every soldier's or Veteran's experiences the alienation that is central to *The Forever War* is, perhaps counterintuitively, still an excellent literary statement of some experiences of many of those individuals. *The Forever War* "is an extended metaphor on Vietnam," Haldeman has claimed, "mainly about the alienation of soldiers and veterans from the culture they risked their lives to protect" (1993, 146). In using both the tropes and clichés of American science fiction, with its specific mythological underpinnings of the frontier experience, Haldeman's statement on alienation cannot be read as anything other than a rejection of that mythological construct. Perhaps because, as Tim Blackmore has claimed, "Vietnam provides literal, metaphorical, allegorical, and anagogical ground for the discussion of the alien," Haldeman's *Forever War* is primarily concerned with the great depth of that alienation (1993, 135). Insofar as Mark Schorer's definition that myths "are the instruments by which we continually struggle to make our experience intelligible to ourselves" operates throughout this work, Haldeman's inversion of the tropes of science fiction, based as they are upon frontier myths, is a statement that those frontier myths no longer have the capacity to fulfill Schorer's definition (1968, 355).

William H. McNeill, in his essay on "The Care and Repair of Public Myth," has argued, "myth makers and myth breakers are entrusted with the task of adjusting and improving received ways of understanding and reacting to the world" (1986, 35). Haldeman's *Forever War* is an exercise in McNeill's "myth breaking" in that it argues that the "discrepancies between old myths and current realities are great enough to be troubling" (1986, 30). For Haldeman, like a great number of other writers, the Vietnam War represented an outright negation of the myths of the frontier that had previously "ways of understanding" the American experience in the world. In as much as those myths had been the basis for many of the tropes of American science fiction, in using and rejecting those tropes Haldeman was also commenting on the myths that proceeded—and produced—them.

American rhetoric scholar Madena Bridges Creek has examined the ways in which literary and popular nonfiction and fiction writing about the Vietnam War reacted to—and often against—the prevailing American mythos that existed prior to the conflict. Her 1982 doctoral dissertation examines some of the classic American Studies works considered here, and her conclusion about them is also similar: American Studies scholarship views that mythos as "collective ways of viewing reality based as much upon subconscious need as upon empirical evidence" (1982, 31). For Creek's examination of the very realistic representations in both fiction and nonfiction about Vietnam offered "an extreme experience, an experience outside the bounds of everyday reality" that threatened the validity of that American mythos because its "inadequacies [became] appear ant to those basing their identity on them" (1982, 31). It is a key component in Creek's work that she is examining what are essentially meant as works of literary realism—works similar to Haldeman's *War Year* and *1968*. When a novel, such as *The Forever War*, however, works so hard to present alienation as its primary theme and does so through the nonrealistic genre conventions of science fiction it is inevitably fated to reject myth—both the Malinowskian "societal glue" and what Schorer called myth's "organizing value for experience" in a far more generalized sense. More specifically, where Creek found in the realistic portrayals of Vietnam works that exposed the "antinomian side of the American character and . . . the unresolved contradictions at the heart of the American experience," Haldeman's *Forever War* offers more than merely another round of ammunition on one side of that ongoing debate (1982, 168). Rather, by using the tropes of science fiction to undermine the mythos of both that genre and Creek's "American experience" on which it is built, Haldeman offers not merely another work about the experience of Vietnam but is also a work that rejects the Vietnam War and the mythos that is perceived to have led to it. Furthermore, because of the theme of alienation at the center of the work as well, it seems to reject myths—all myths, not merely American ones—as well.

But for Haldeman that alienation was somewhat more spatially oriented. In an essay, written much later, called "Vietnam and other Alien Worlds," Haldeman pays particular attention to his arrival "in country" and, specifically, in another hemisphere. A sky-watcher from childhood and the holder of a degree in astronomy and physics, Haldeman's reaction to his arrival in Vietnam was a reaction made more specific by his first viewing of the sky of the southern hemisphere. "It's an eighteen-hour flight from the West Coast to Vietnam, and we were about an hour from landing when I first had the sense of entering an alien world," he wrote (1993, 139). But it was not the jungle, or the military presence, or even the foreign culture that offered the sense of alienation. It was the stars:

> I couldn't recognize any of the stars. That was profound: I had been interested in astronomy, sometimes to the point of fanaticism, since I was a Cub Scout, and the constellations were as familiar to me as the faces of family members . . . For the first time in my life I was looking at the Southern Cross, Alpha Centauri, the Magellanic Clouds. I stared for a long time, transfixed, as the plane began its descent. (1993, 139)

Landing in Vietnam, for Haldeman, was indeed landing in an alien world. The base he arrived at—a sprawling base miles in length in South Vietnam—was under attack when he got there. He was not asked to join the fight, which was taking place at the other end of the base, but to serve food in a mess hall. His first day in Vietnam, then, he "served instant mashed potatoes to hundreds of people who didn't seem to notice there was a war going on outside." Such ironies would later be found in his writing. His first novel, *War Year*, published just a short three years after his return from the war, would begin almost exactly as did his own experience in Vietnam. The very first lines recount the same experience, with only slightly different details: "I almost slept through that first enemy attack. I'd been on KP [Kitchen Police] all day, washing dishes, on my feet from dawn to dark. . . . I didn't hear the sirens when they went off. I woke up with this big guy shaking me" (Haldeman 2005f, 5; originally published as a short novel in 1972). But where the nineteen-year-old protagonist of *War Year* is best described as a stereotype of that age, the twenty-four-year-old Haldeman was somewhat more reflexive. And years later, in a more personal fashion, Haldeman could describe the experience of serving KP in a mess hall while the base it was in was under attack. The incongruity of the situation—and possibly the irony of much of the American experience in Vietnam itself—could come through more clearly. For after his shift serving mess, on his first day "in country," Haldeman did what many soldiers would do: he slipped away for some rest. Rather than dwell on the attack on the base, or the level of fear a new Army soldier experiences near a firefight, Haldeman simply experienced the alienation of the situation. A committed stargazer and astronomy student, the soldier's slang of referring to home as "the World" would be yet another irony that certainly fit Haldeman's mind-set. He was quite clearly no longer in the World. Later he would recall it this way:

> After my shift I spent a couple of hours sitting behind the mess hall, sipping warm beer and watching the helicopters bob and weave, spitting flame, firing rockets. Men dying under alien stars, literally. (1993, 139)

Men dying under alien stars, alienated from the society they had been sent to protect, is a key theme of *The Forever War*. But to understand precisely how Haldeman's science fictional version of Vietnam rejects both the experience of that war, and the mythology that the war seemed to have grown from—and perhaps myth generally—it is necessary first to understand the profound effects the war had on American culture more broadly. The Vietnam War was transformative in that it altered significant portions of American culture and picayune in that it destroyed the Myth and Symbol school of American Studies. It was transformative in that it reshaped the American political landscape and trivial in that those changes left schools of academic inquiry forever altered. It remains, even now with additional and ongoing wars more recently in the forefront of American national consciousness, the most significantly affecting conflict since the American Civil War. And, like the Civil War, the Vietnam War presented an affront to the defining mythology of the nation and to the defining way of understanding that mythology used by scholars. In a way the Vietnam War first splintered the American mythos, and *The Forever War* dealt with the consequences of that disintegration. Thus, to understand Haldeman's novel it is necessary first to understand the mythical implications of the Vietnam War, and to understand both is to understand the both the profound and the subtle mythological reactions that would follow in American culture.

THE TET OFFENSIVE, THE MY LAI MASSACRE, AND AMERICAN MYTH

The ramifications resulting from Vietnam War on American myths has consumed entire monographs. After all, the war took the lives of roughly sixty thousand U.S. servicemen and women and possibly in excess of two million Vietnamese.[1] The disparity between these figures, however, has not led to a similar disparity of viewpoints among American scholars. H. Bruce Franklin has described a situation that demands a certain amount of caution:

> Within the dominant American culture, "Vietnam" is no longer a nation, a people, or even a war. "Vietnam" is something terrible that happened to us, something that divided, wounded, and victimized America. (2007, 34)

Franklin's point is twofold, and only one aspect applies to this work. For American historians, political scientists, other scholars and public intellectuals the dominant theme to grow out of the Vietnam War has had far too little to do with the actual Vietnamese, and much to do with Americans. While this chapter is more concerned with how the experience of Americans in the conflict moved the nation toward Franklin's "divided, wounded" state, it remains a point of intense importance to remember that, in actuality, the events in Southeast Asia were far more deleterious for the Vietnamese, Laotians, and Cambodians than any others. As a work analyzing the American mythscape that was significantly altered in light of the Vietnam War, it is necessary to emphasize the *American* side of this very unequal equation. Yet it should not be forgotten how truly unequal that equation remains.

Any discussion of the American side of that equation—that is, of the Vietnam War and its implications on American mythology—can be addressed with an examination of some of the more pointed examples of the conflict. Among those possible touchstones a number of events could easily be used. A starting point for such examples could be the 2 August 1964 Gulf of Tonkin "incident," in which the U.S. Navy destroyer USS *Maddox* allegedly came under enemy fire from North Vietnamese torpedo boats, thus offering President Lyndon B. Johnson a justification in asking Congress for the authority to deploy American troops in Vietnam. And, on April 29–30, 1975, as North Vietnamese regulars entered Saigon and Americans and their

Vietnamese allies were evacuated from the American Embassy compound by helicopters with many of those very same helicopters eventually being pushed off the flight deck of the USS *Midway* aircraft carrier to make room for more refugees, could offer an excellent end point of such examples.

Between those historical two points, however, there are two other events that offer compelling reasons to focus on them: the Tet Offensive, beginning at on January 31, 1968, and the events at the My Lai and My Khe hamlets in the village of Son My on March 16, 1968. Each stands out for its notoriety and for its broad effects throughout American culture. While Tet comprised a nationwide attempt by the North Vietnamese to engender an active rebellion in South Vietnam, and the events now known as the My Lai Massacre saw the deaths of something approaching five hundred or more Vietnamese civilians in a relatively small area, the two events together can be seen as major mythological turning points of the American involvement in Vietnam. Indeed, both separately and when combined the events work as metonyms for all of the military, political and mythic failures of the Vietnam War.

The Tet Offensive, in itself, was an absolute failure militarily for North Vietnam and their Viet Cong allies (cf. Anderson 2006; Schmitz 2005). The attempt to incite active rebellions did not happen and American troops and their South Vietnamese allies responded quickly and, for the most part, effectively. In the opening hours—and then days—of the Offensive, however, this did not seem to be the case. Five of the six largest cities in South Vietnam were attacked almost simultaneously, as were thirty-six provincial capitals and nearly every military base. Vietnam's old imperial capital city of Hue was captured and held for weeks by the North. As the conservative American writer Michael Lind, whose work on the war is provocatively subtitled *The Necessary War*, could describe it, however, the "failure of the Tet Offensive was a devastating military setback for the communists . . . The war was [still] winnable—but it was far from being won at a cost acceptable to the American public" (1999, 21). In mythological terms the Tet Offensive was a defeat in that it brought with it a surety of knowledge that the war the United States had been officially engaged in for three years was not, as had been communicated to the American public, a war between a small number of technologically backward communist insurgents, their foreign communist supporters, and an overwhelming, technologically advanced American fighting force. Regardless of whether Lind is correct that the war was still winnable any possible victory did not seem, as it had before, inevitable. Rather, the broader message taken from the Tet Offensive, as historian Allen J. Matusow has written, was that "[i]t showed that no place in South Vietnam was secure, not even the American embassy, whose walls had been breached in the first hours of the attack" (1984, 391). Therefore, though the forces of the United States and their South Vietnamese allies were, eventually, victorious and inflicted incredibly heavy casualties on their attackers, Matusow is describing a far more serious point when he notes that when commanding "General William Westmoreland . . . issued victory statements after the Offensive ended, few took them seriously though militarily they were right" (1984, 391).

Thus, while public opinion did not demonstratively change immediately following Tet, the Offensive did begin the process in which it became possible for the majority of Americans to consider the war problematic—and not for military reasons as much as for ideological ones. While support for the war has already been "eroding for the past year," as David F. Schmitz described it, Tet "continued that process as American objectives and national interest in Vietnam became more unclear all the time" (2005, 158). Further, as Daniel C. Hallin has reported, media coverage in the United States began to decisively turn against the war, with editorial comments on television news, for example, somewhat suddenly running fully "two to one against" the war after Tet (1986, 161). This lack of clarity and change in media coverage was largely a failure of the military events of Tet to match the prevailing mythological structures available to the American public, and the mythological breach was far more significant a defeat that the American military's battlefield victories could ever overcome. It was not so much that the offensive was disastrous to the military forces aligned in South Vietnam as it was that the offensive fundamentally altered the image of the American military and moved it decisively away from the actively projected ideal of a traditional battlefield combat invincibility.

With Tet the notion that the United States was engaged in a conventional battlefield-based war with associated notions of limited civilian casualties, superiority of military and technological might, and easily quantifiable "victories" based on casualty reports was fully undermined. Those notions were also inextricably combined with another set of mythical images—those of moral certitude in the American prosecution of its war aims. The American journalist Frances FitzGerald explained the immediate issues for the American public that resulted from Tet succinctly. "Over the past year" American military leaders had, FitzGerald argued, "had given the impression that the enemy threat had receded, that the American troops had pushed the main force units out of the populated areas into the jungles of the border" (1972, 489). Tet contradicted such previous reports and, as a result, contradicted the image the American military was trying to project back home from the war. As a result, as H. Bruce Franklin has described it, the Tet Offensive "certainly succeeded in convincing the vast majority of the American people that victory for the United States was not a feasible outcome of the war" (1993, 43). But it also fully set the stage for the acceptance of not just a battlefield quandary in Vietnam but of a moral quandary as well. More specifically, if the Tet Offensive convinced Americans that victory was not something they could naturally presume, it also made possible that the events at My Lai, which would become public knowledge in 1969, would add another aspect to the attack on the dominant American mythology: not only might the American side lose, but no longer were the Americans even the "good guys."

A potential defeat in a foreign war was, to the American public, merely the first stage in an unprecedented cultural subversion of American mythological structures. As a result of Tet, "the aftershocks" as H. Bruce Franklin put it, would soon be "rocking American society" (Franklin 1993, 43–44). A significant part of those aftershocks were a readiness to accept further bad news about the war by a majority of Americans. In other terms, one effect of the Tet Offensive was a heightened willingness to accept, later, the reporting of the war crimes that had taken place in the hamlets of My Lai and My Khe. British historian of U.S. history Kendrick Oliver has noted that "the events of Tet helped to create a space within [American] culture of war reporting that would eventually permit the revelations about the My Lai massacre to be believed, registered as significant and published as news" (Oliver 2006, 29).

My Lai, an unqualified war crime in which U.S. soldiers participated in the illegal killings of possibly in excess of five hundred civilians in the hamlets of My Lai and My Khe has acted, according to Richard Slotkin, "as a metaphor for what had gone wrong with the war" as a whole (1993, 587). More than simply that, however, the My Lai massacre occupies a mythological rupture point in American society.[2] More than simply a metaphor for what had gone wrong, however, the events at My Lai also act as a metonym for a host of related atrocities committed by the American military during the war. FitzGerald described the events in a brief overview that captures not just the specifics but also sets up precisely how the massacre will become a threat for American myth:

> On a routine search and destroy operation a company from the American Division had walked into the village of My Lai and without provocation had gunned down three hundred and forty-seven civilians, most of them women and children. A photographer had taken pictures of screaming women, dead babies, and a mass of bodies piled up in a ditch. Even once substantiated, the story seemed incredible to many people. How could American soldiers have committed such an atrocity? (1972, 463; the specific number of Vietnamese casualties varies by report, but more recent scholarship suggests FitzGerald's numbers are on the low end of estimates.)

The question asked by FitzGerald is a question concerned with the mythical underpinnings of the American conception of its fighting force as much as it is concerned with anything else. The atrocity committed by American soldiers not only broke the army's own rules of engagement, but they failed to meet the requisite "good guy" standards of American myth. As Kendrick Oliver has put it, as a result "what Americans thought about the massacre, then, was never really to cohere into consensus, into the sort of stable and accepted 'truth' most easily converted into a usable past" (2006, 9).

Perhaps even more importantly, however, the events at My Lai and My Khe were, as Frances FitzGerald described them, "not exceptional to the

American war" (1972, 463). The massacre was, instead, merely the best-known example of those times in which American soldiers behaved abhorrently. The fact that many other American soldiers did not behave in such a manner—and, actually, many behaved admirably—does not alter the threat to American myth presented by My Lai. And, when combined with the convergent mythic threat of the Tet Offensive, the two events offer a combined metonymic signifier of the failures of American mythology in the jungles and rice paddies of South East Asia.

Precisely how these two events combined to undermine American myth is a complex process and one open to serious debate of form and method of analysis. However, John Hellmann has offered an analysis of the significance of Tet while Richard Slotkin has discussed the massacre at My Lai, and both have done so in ways that both summarizes the mythological implications of those two events while also offering a set of related connotations. Hellman's examination is largely confined to the primarily negative reviews of the 1968 John Wayne film *The Green Berets* while Slotkin looks to the photojournalism of *Life* magazine (Kellogg and Wayne 1968; Wingo 1969). Together these two reviews offer important insights into the Vietnam experience as a whole.

For Hellmann, sixty-year-old John Wayne's portrayal of Green Beret Colonel Mike Kirby offers an attempt to "ludicrously [impose] the world of the Hollywood western and war film—the American mythic landscape—upon the Vietnam setting" (1986, 91). The film can best be described as a John Wayne vehicle designed to present the Vietnam War as being exactly what Hellman offers, placing it firmly in line with previous tropes of Wayne films such as those westerns and World War II films that made him famous. The film looks little like Vietnam but rather like the area around Fort Benning, Georgia, where it was filmed. And, at a time when the actual images coming from Vietnam in the wake of Tet did not, as Hellman put it, "correspond to the roles that the national myth ascribed to the American, his noble native ally, and their mutual enemy the savage or European aggressor," *The Green Berets* worked hard to correct issue (1986, 89). Immediately after the Tet Offensive Americans were "left without a convincing story of the conflict in Vietnam" that matched the narratives implied by American myths. (ibid., 90). Wayne thus offered one, and it is a narrative built more on the various tropes of his own acting past than on the actual events in the war. Critically rebuked, in part because of its conservative politics, it also had the problem that Wayne played, essentially, what was "hardly a character at all [but] the John Wayne persona [that] has been reduced to a simpleminded parody of itself" (Anderegg 1991, 25). At the same time what Wayne's movie provided, in Hellman's analysis, was an image "showing Americans their preferred self-image." The film thus offered

> a small band of rugged yet pure-hearted individualists, on a frontier landscape, aiding pastoral natives against both wild savages emerging

> from the anarchic forest and robot-like soldiers extending the oppression of their machine-like society. (92)[3]

Wayne could even end the film—a film he codirected, after all—with a nearly impossible image of his character leading a Vietnamese child, hand in hand, into what appears to be a western sun setting into a coastal beach. When the boys asks Wayne's Colonel Kirby what he is to do now that his favorite Green Beret has been killed in action, Kirby replies with all of the paternalistic tropes his long Hollywood career provided. Placing the dead soldier's beret on the Vietnamese boy's head Wayne proclaims, "You let me worry about that, Green Beret. You're what this thing's all about."

But taking care of a young orphaned child and making him an honorary Green Beret in the process was not what the American involvement in Vietnam was all about at all. And while Hellmann argues that the overwhelmingly negative critical reactions to the film "both reveal the impact of the antiwar movement's story upon public consciousness, and represent a powerful new recounting of it," it is also a statement about the status of the very myths Wayne offered as well (ibid., 92) The film might have been rejected by the top critics but it still made at the domestic box office "close to $20 million," and given "ticket prices in 1968, that was a brisk tally," especially when one considers that the film cost only $6.1 million to produce (Devine 1999, 45). As Richard Slotkin offered in his own critique of the film, the rejection of *The Green Berets* by major news outlets at the time were as much about Wayne's offering of the traditional mythic tropes as they were about "the fidelity with which he represents the officially authorized version of events" in Vietnam (1993, 532). Yet Slotkin, Hellman, and other scholars are not quite comfortable with the substantial profits—and thus the substantial audience—for the film. For Slotkin *The Green Berets* is simply part of a larger trend of "symbolic iconography of a growing uneasiness about the ways in which the American military effort was affecting the 'little people' it was our 'mission' to rescue." And that "growing uneasiness" would move from Wayne's film to the "culminating [images] in *Life* magazine's 1969 publication of the pictures of the [My Lai] massacre" (ibid., 532).

Hellmann's analysis of Wayne's *The Green Berets* is specifically directed toward a cultural response to the Tet Offensive while Slotkin's analysis of My Lai is merely directed toward that horrendous massacre. However, both scholars' conclusions apply equally and, perhaps appropriately, to the Vietnam War as a whole. Both scholars also clearly intend this to be the case. In part this is because Tet offered a mythological precursor that allowed the negative reception of the My Lai massacre to be possible. As a result these two events exist on a continuum of cultural reactions in the period. More fully, however, this is because Hellmann's analysis of Tet through *The Green Berets* and Slotkin's analysis of My Lai are each an attempt to assess the mythographic aftermath not just of those events as they are intended as representative of the entirety of the Vietnam War experience.

Thus, superficially at least, Slotkin's analysis may appear to be an eclectic method of inquiry, but in reality it offers an excellent approach for examining the ways in which the serious blows from both Tet and My Lai were struck against American mythology. For Slotkin the *Life* magazine coverage offered a narrative in text and images that allowed readers to "see" My Lai as a cinematic spectacle that transformed "a disruptive military/political event into a crisis of American mythology" (ibid., 581). Many of the most famous and most gruesome photographs from My Lai first appeared in *Life*, and for Slotkin those images and the accompanying text offer a very specific narrative. That story begins with American soldiers as the standard image of heroic warriors and progresses to a point in which "instead of protecting women and children from 'the horror,' they have themselves *become* 'the horror'" (ibid., 585; emphasis in the original). This is fully an inversion of the traditional mythological format, making My Lai a "counter myth" in which the "inversion of the normal war-movie/Western scenario is now complete" (ibid., 586). That "war-movie/Western scenario"—which, importantly, is very much a part of Wayne's *The Green Berets* as well—becomes symbolic of a situation where the archetypal American heroes have become, mythologically, the villains.

The essential point made by Hellmann and Slotkin, however, is that this inversion of mythological types is a paramount result of the American experience in Vietnam. At the same time, then, it is an odd fact that so much of the literary response to the war—written by journalists and veterans alike—work at presenting the war within a defined mythological framework. Most often this is in the form of the huntsman entering the frontier, of the leaving behind of civilization, and of the encounter with the alien "other" or the savage. And perhaps even odder is the fact that the critical responses to such works also endeavor to place them within a broad framework of American literature. Thus, the literature resulting from Vietnam became, for the critics who examined it, merely another example of the American literary tradition of the frontier.

It is in not being able to be placed among such a literary tradition—while still being very much aware of it—that Joe Haldeman's *The Forever War* excels. It excels as a novel of Vietnam, of science fiction, and of the unmasking of the conventions of mythic frontier narratives in both of those traditions.

THE FOREVER WAR: DYING UNDER ALIEN STARS

Literature scholar Philip H. Melling noted that the literature that grew out of the American experience in Vietnam had to confront a variety of cultural inconsistencies. As a result an "emphasis on crisis and contradiction and the retreat from empiricism," Melling claimed, "are the tithes the Vietnam writer pays the critic if he wishes to be seen as worthy of relevance" (1990, 7).

Although Melling disagrees with such postmodern interpretations, he does note that writing about the war—be it a straight autobiography, a more literary-minded memoir, reportage, or fiction—have all avoided "the temptation to seek coherence" in the way the experience of the war is represented (ibid., 5). But in their attempts to fashion something of artistic merit from the ashes of the American experience in Vietnam writers were forced to confront Melling's "crisis and contradiction." To do so required attempts at confronting war, itself, often in new and experimental ways. American literature scholar Jerome Klinkowitz put this bluntly:

> Because of the radically different nature of this war, both in its initial combat abroad and in the resistance to it at home, conventional modes of understanding were discarded by many as obsolete. That was the first theme announced by the earliest novels to come out of the Vietnam experience. (1980, 76)

For Klinkowitz, however, those new ways of writing about war that were birthed by Vietnam are all about an "aesthetic situation" that "reverses so many traditional values" (ibid, 81). For a scholar such as Klinkowitz—who made his career through examinations of experimental fiction by the likes of Joseph Heller and, most extensively, Kurt Vonnegut—the American reaction to Vietnam, for all its tendency to highlight "experiments" in narrative, amounted to little more than the realization that American "myth had been wrong for centuries until Vietnam revealed what it really was" (ibid., 87).

The literary realization that the Vietnam War was, somehow, different has led to both profoundly new literary works and, at the same time, profoundly narrowly focused critical analyses of those works. Thus, the published products coming out of the Vietnam experience offers, as Jon Roper has put it, a "range and diversity of writing from those with personal experience of [Vietnam]" but critical approaches have lacked diversity in their selection of texts (2007, 13). Another American literary scholar who has written on the works coming out of Vietnam, Philip D. Beidler, could rightly claim in the introduction to his *American Literature and the Experience of Vietnam*:

> Most of the time in Vietnam, there were some things that seemed just too terrible and strange to be true and others that were just too terrible and true to be strange. Indeed, the dominant impression one got of the American experience there . . . was of pure, manic contradiction, of the abiding unreality, in some cases, of that which seemed most real, and of the utter reality, in others, of things so monstrously concrete and immediate that they could only be handled by imaginative conversion into some unreal other. (1982, 4)

But in offering such a summation Beidler could only offer analyses of a surprisingly noncontradictory group of works best described by the moniker of "traditional." So while he could offer what he called part of "the larger process in cultural myth-making," the vast majority of his texts fit squarely within what might be called "mainstream literature" and were published by major New York publishers (ibid., xi). And while he could admit that "Vietnam would always be a place with no real points of reference, then *or* now," he offers no analyses of works that do not, at least in some way, attempt to attach their narratives to such familiar points of reference (ibid., 16; emphasis in the original). Indeed, Beidler eventually devotes an entire volume to the idea that "even the desperate literary enterprise amidst the competing claims of history and myth" born out of Vietnam is best understood when compared to similar works emerging from other wars, especially World War One (1991, 3). Thus, while he might claim his undertaking to be an examination of the "continuing work across a vast range of genre and mode" of various Vietnam writers, his final conclusion on that writing remains that for "all their terrible, urgent strangeness, they also possess an older, almost ritualized quality of iconographic permanence" (Beidler 1991, 5; Beidler 1982, 25). Beidler's is not a thesis designed to highlight the newness of Vietnam writing or to truly examine a "vast range of genre and mode" but to contain it within traditional boundaries. Here he differs little from other critics. Jerome Klinkowitz offers a similar disconnect, arguing of the literature of the period that "when the [Vietnam] war began, new strategies were needed to fight it, new techniques to describe it, and a new mentality to understand it" (qtd. in Beidler 1982, 75–76). And yet, despite this assertion Klinkowitz, too, ends by inserting the new literature derived from the Vietnam experience into a traditional order claiming that "in aesthetic terms," the literature and other art forms of the period "are the climax of trends active since the beginning of the century" (ibid., 102). Another Vietnam veteran and American literature scholar, Toby C. Herzog, could thus offer an examination of works that "all realistically and perceptively define the age, the war, and the people who participated in the Vietnam experience," and end up examining many of the same texts that other scholars have already offered. As a result he can only take "Vietnam war stories" and analyze them "by connecting them to historical events, other modern war literature, and American culture" (1992, 10).

In short, while academic critics have been swift to point out those aspects of Vietnam literature that are revolutionary in their newness in the process of canon creation, most have also succumbed to Philip Melling's "temptation to seek coherence." For while the writers of the Vietnam experience avoided that temptation, their critics waste little time in placing this new writing within the confines of traditional categories, offering up their texts as merely modifications of existing traditional forms. Melling can offer a genealogy of Vietnam literature that is traced back to Puritan narratives in Colonial New England, Klinkowitz can begin his analysis by quoting Thomas Jefferson, and Beidler can declare without irony that Vietnam literature "often turns out to be very much in context, so to speak, with regard to our national traditions of literature and popular myth-making" (1982, 19). So while the actual writing coming from the Vietnam experience might, as the American scholar James C. Wilson put it, "take to an extreme the unreality, the discontinuity, and the loss of values that may characterize much of our experience in America today," the critical response to that writing has been far more concerned with established canons and a "temptation to seek coherence" (1982, 102). It is not surprising, then, that Melling, Beidler, Klinkowitz, and a significant number of other scholarly writers end up writing about the same texts by Philip Caputo, David Halberstam, Tim O'Brien, Robert Stone, Michael Herr, and others. In part, truthfully, the canon of Vietnam literature was being created through these early critical works. But, at the same time, that canon was created primarily to mesh with an existing, long-established canon of American letters generally, and as a result, there is an unsurprising tendency to marginalize works outside of that established tradition. What is surprising, however, is how much critics such as Beidler have to say about works that do fall out of that tradition—even when they fail to specifically acknowledge them. Much of what Beidler, Klinkowitz, or others have offered about the "newness" of Vietnam literature—even when they eventually undermine those statements, themselves—are quite usefully applied to works that fall into genres that they, as academic literature scholars, usually ignore outright.

MOVING BEYOND THE CANON

This tendency to ignore works outside of "mainstream" literary studies is an issue that H. Bruce Franklin has addressed in some detail. In his *Vietnam and Other American Fantasies* Franklin attempts to move beyond "traditional canons" and assesses science fiction, comic books, and other "alternative" texts surrounding the Vietnam War. Taking to task Beidler, Hellman, and a number of other scholars in a lengthy expository endnote when writing about Haldeman's *Forever War*, Franklin noted the novel is "not even listed" in many extensive bibliographies and sardonically quips that Beidler "does give half a sentence to Haldeman's . . . *War Year*" (2000, 226n38). Franklin's conclusion is that *The Forever War* should be taken more seriously than it has been. However, although Franklin frequently references the novel in advancing his own theses—particularly on fantasy representations of American militarism and the American penchant for proposing superweapons—he, too, has never offered a sustained reading of the novel (Franklin 2000, 2008). Yet this is precisely what Franklin calls for in arguing that criticism of Vietnam writing has followed an overly predictable and, as a result, narrow path.

Indeed, for Franklin a major issue with the critical response to writing on Vietnam lies in its traditionalism. A need for an examination of genre writing is central to his work and, he believes, a necessity for any critical response to Vietnam writing. "America's war in Indochina," he has claimed, "cannot be separated from American science fiction" (2000, 151). Thus, the broad consensus among scholarly critics to discount short fiction and novels by science fiction writers, as well as in other media, remains a major flaw in academic discourse about the war. He makes his claim quite bluntly:

> Out of American pulp, comic book, and movie science fiction of the 1930s, 1940s, and 1950s poured two streams of images that profoundly influenced how the [Vietnam] war was conceived and conducted: fantasies of techno-wonders and of superheroes . . . Indeed, the war cannot be fully comprehended unless it is seen in part as a form of American science fiction and fantasy.
>
> (ibid., 151)

For Franklin a cultural understanding of the experience of Vietnam requires an understanding that the war "has been transformed into a direct source of fantasies central to American culture" and this means an examination of the cultural artifacts resulting from the war must include texts well outside the mainstream of the literary intelligentsia (ibid., 172). So while Klinkowitz might have hoped for works outside of the "conventional modes of understanding," and Beidler could recognize the need for writing that addressed Vietnam as a place "with no real points of reference," Franklin turned to a number of texts that offered exactly what other literature scholars claimed to be seeking but had, somehow, missed or simply did not want to acknowledge. And chief among these texts must be Joe Haldeman's *The Forever War*. For H. Bruce Franklin the work is

> perhaps the biggest best-seller of all novels about the Vietnam War. Despite widespread admiration among science-fiction critics and sales of well over a million copies, *The Forever War* is generally ignored by surveys and criticism of Vietnam literature.
>
> (ibid., 164)

In comparing *Forever War* to the types of works scholars such as Melling, Klinkowitz, and Beidler have examined Tim Blackmore concludes that "Haldeman's grasp and manipulation of American myth puts him in a class above other 'Vietnam' writers," and then goes on to list the writers he means (1993, 131). Blackmore's list reads as though it is a table of contents of the usual suspects of the established and standard works on the literary response to Vietnam. For Blackmore, Haldeman succeeds where writers such as Michael Herr, Philip Caputo, and Tim O'Brien, for all their merits, falter. Haldeman offers "various forms of thematic, narrative, and stylistic inversion," allowing him to offer a critical perspective on those very themes, narrative forms, and styles in the guise of science fiction (ibid., 133).

By offering a seriously playful use of the well-worn themes of science fiction, Haldeman can use those themes for purposes that might seem counter-intuitive. Additionally and perhaps more importantly than the sales figures for *The Forever War* or the fact that the work actually provides the kind of insight many scholarly critics claim to be looking for, the novel can also offer an analysis from the one genre of American fiction that specializes in extrapolating the effects of technology. This is especially important in a war that, as Toby Herzog put it, "ended American innocence about . . . the effectiveness of technology, and battlefield brutality" (1992, 214). A war that was a technophile's dream, fought while Americans also planned their lunar landings, requires insights from a genre not afraid of such technology. As Vietnam veteran William Broyles, Jr., once described it:

> Other Americans were on the moon, sent there by the same American can-do spirit that had sent us to Vietnam, and using the same technology that was so useless there. Those astronauts were the first men to leave the earth for another celestial body, but they knew more about the moon than we knew about Vietnam. We were strangers on our own planet, alone and afraid. (1996, 11)

Haldeman, who had described the war as "dying under alien stars" after all, uses the full range of science fiction conventions to discuss the Vietnam War. Not afraid of working through that "American can-do spirit" to its logical ends, or of extrapolating fictionally about the technology future wars might use, Haldeman is able to offer Beidler's "things so monstrously concrete and immediate that they could only be handled by imaginative conversion into some unreal other." In doing so Haldeman undermines all of the science fiction conventions at his disposal at every opportunity, always to further his narrative. In using those conventions in *The Forever War* Haldeman enters into what another American literature scholar, Thomas Myers, has called the "battle for collective memory" about the Vietnam experience. For Myers that is "an ongoing pitched battle for the correlation of new emotional, psychological and spiritual data within larger mythic narratives [that include] the production of texts that are . . . from a limited historiographical perspective perhaps, less trustworthy" but, at the same time, are more truthful (1988, 5–6).

Some of the conventions Haldeman employs are so widely used that they cannot be considered actual conventions, or even tropes, so much as clichés. But faster-than-light travel, interstellar war, the technological advances of weapons, and, most important, expansive frontiers are all central to American science fiction and to *The Forever War*. David Mogen notes this connection, and the significance of it: "Though science fiction frontiers have often been regarded as artificial transplants of the past on to the future," he has written, "they are more accurately understood as products of a cultural mythology" (1993, 7). For Mogen the tropes and conventions used by

American science fiction writers, based on frontier mythology, are simply part of the "genetic code implanted in the American imagination" (ibid., 7). Yet Haldeman repeatedly takes these established conventions, ironically inverts them for his own narrative's needs, and then uses them to comment on the Vietnam War.

Franklin, in praising such a narrative strategy, declared that it seems Haldeman "delights in twisting the futuristic hardware and adventure formulas of old-fashioned militaristic science fiction into their opposite" (2008, 194). This particular narrative strategy—the use of common tropes for counter-intuitive purposes—enables Haldeman to establish his novel as at once traditional science fiction and at the same time very revolutionary science fiction. Franklin's "delights in twisting" such formulas, or what Blackmore, also using that word *delight*, called Haldeman's "thematic, narrative, and stylistic inversion," allowed Haldeman to move beyond simply co-opting the conventions of science fiction. He is actively critiquing the frontier mythology upon which those conventions are based. An examination of each of the "stylistic inversion" of tropes, and how Haldeman accomplishes this serves not only as an examination of the novel but also as an examination of the effects of the Vietnam experience on American mythology as well.

Written while the Vietnam War was still being fought, *The Forever War* is both about what was until that time the longest war in U.S. history and a science fiction novel commenting on war holistically. Begun in 1971, serialized in the science fiction magazine *Analog* in 1972 and 1973, and published as a novel in 1974, the novel moves through both the height of the war in Vietnam and attempts at reintegration after war that were the experience of American society in the same period (Gordon [1980, 25–26] offers a detailed publication history of the novel as short stories and Haldeman's work with the text throughout the period). Although Joe Haldeman has been quick to point out that neither of the protagonists in his two realistic Vietnam novels was autobiographical, his science fiction version of that war is narrated by a protagonist remarkably like Haldeman himself. William Mandella, a graduate with a physics degree, is drafted via the "Elite Conscription Act of 1996" and, in the nonrealism of science fiction, serves the entire span of "the Forever War"—a "1143-year-long war [that began] on false pretenses and only continued because the two races were unable to communicate" (Haldeman 2004c, 249). Even the protagonist, William Mandella, it turns out, bears Haldeman's middle name and a near anagram of his last name (Franklin 2000, 165). Many of the letters written from Vietnam by Haldeman were to his eventual wife, Mary Gay Potter, and similarly, Mandella's major love interest in *The Forever War* is MaryGay Potter. And, like Haldeman, Mandella finds himself in battles that seem, at best, unorganized chaos. Resultantly, Franklin has called *The Forever War* "a kind of autobiography in fantastic disguise" (2000, 165).

A novel of a war between Earth and an incomprehensible alien enemy, the Taurans, *The Forever War* utilizes many of science fiction's most well-worn clichés. In it a squad of future soldiers, operating from faster-than-light starships, fight to protect humanity and their native planet. The military technology of the future is described in detail. As is how soldiers use that technology to fight, the physics of interstellar travel, the effects of such travel on the passage of time for the characters, and social change on Earth. The novel that *Forever War* is most often compared is Robert A. Heinlein's 1959 military science fiction novel *Starship Troopers*, which the American cultural historian De Witt Douglas Kilgore accurately described as a work of "glorious militarism" (Heinlein 2005; Kilgore 2003, 190). The two novels share a superficial resemblance as both offer a first-person narrative of a soldier through entrance into military service, basic training, various battles, and eventually training as an officer. Both also utilize the trope of technologically advanced powered exoskeleton suits for battle. Haldeman said as early as 1977 that he did not consider the Heinlein novel until he had written fully a third of *Forever War*—a statement he repeated to me, using almost the exact language, more than thirty-five years later (Schweitzer 1977; Haldeman 2011b). Yet, in science fiction circles, the novels continue to be compared. *Starship Troopers*, however, is a work that both uses established science fiction themes of militarism on the frontier and further establishes them within the genre. As the American science fiction writer and critic Alexi Panshin described it, *Starship Troopers* presents all "the soldiers we see" as "tough, smart, competent, cleancut, clean shaven, and noble" (1968, 96). It is a work that, as H. Bruce Franklin noted, "displays the superelite force designed to fight the permanent wars necessary to fulfill Earth's manifest destiny in the galaxy" (1980, 111) The stereotypes of "glorious militarism" are, in fact, everywhere.

David Mogen, in portraying the work of Heinlein generally, nicely captures the author's use of such frontier mythology and ideology. In Heinlein's work, Mogen notes, "though technology changes, conditions that forge the frontier character endure." As a result Heinlein often "explicitly draws parallels between qualities required to survive the Old West and those required in the new frontiers of space" (1993, 47–48). The battle suit of *Starship Troopers* is precisely this kind of technology, and it offers in Heinlein's work what Mogen claims of much of American science fiction, generally:

> [T]o enlist the energies of our frontier heritage to conquer the new wilderness of outer space that technology has opened to us. Frontier rhetoric marshals the expansive, dynamic energies of Manifest Destiny in the service of a new vision. Both Western and science fiction frontiers provide imaginative alternatives to the present.
>
> (ibid., 34)

Thus, while most comparisons between Haldeman's work and Heinlein's *Starship Troopers* struggle with the perceived political differences between the two works, it is with the differences in the presentation of the technologically advanced armored combat suit in each novel that is most striking and most illustrative of the broader cultural shifts brought about by the Vietnam War. Suits in both novels fully encase the soldier; they offer increased protection, greater strength, and faster movement. Heinlein's narrator originally described them in 1959 simply saying that "[s]uited up you look like a big steel gorilla, armed with gorilla-sized weapons" (2005, 88). But for Heinlein the suit remains the science fiction extrapolation of what H. Bruce Franklin called those "fantasies of techno-wonders" or the American "cult of the superweapon" (Franklin 2008, 5). That is, they represent the ongoing technological fetish of American society for a weapon that will "end all wars." For Franklin the United States has been attempting to perfect superweapons since the dawn of the republic, though he traces the concept in great detail, not surprisingly, to the late nineteenth century and the closing of the American frontier.

Importantly for this work, Franklin notes that as "[f]ascinating as they may be as expressions of psychology and culture," such weapons in American imaginative and speculative fiction "are not primarily fantasies at all" (ibid., 5). For Franklin "American innovations in weaponry have fundamentally transformed not only warfare and geopolitical relationships but the human condition" (ibid.). Heinlein's description of his narrator's battle suit is fully in the tradition of the America "cult of the superweapon." Heinlein's armor is described almost lovingly as a technophiles dream:

> Two thousand pounds of it, maybe, in full kit—yet the very first time you are fitted into one you can immediately walk, run, jump, lie down, pick up an egg without breaking it . . . dance a jig . . . and jump right over the house next door and come down to a feather landing. (88)

There is little to learn for Heinlein's narrator, other than the fact that the suit will "match *any* motion you make, exactly—but with great force." Thus, the soldier can jump, "without having to think about it . . . but higher than you can jump in your skin . . . [and] cushion your landing without your having to think about it" (89; emphasis in the original). In case the reader might have missed the point Heinlein's narrator drives it home:

> And *that* is the beauty of a powered suit: you don't have to think about it, you don't have to drive it, fly it, conn it, operate it; you just wear it and it takes its orders directly from your muscles and does for you what your muscles are trying to do. (ibid.; emphasis in the original)

This is precisely Franklin's "fantasies of techno-wonders" and the "cult of the superweapon" that Franklin has traced to the period between the end of the Civil War and the American entry into World War I—that is, to the period in which Frederick Jackson Turner announced the closing of the frontier. While Franklin does not direct his analysis toward *Starship Troopers* in any great detail, what he says about late-nineteenth- and early-twentieth-century fiction applies nicely to Heinlein's 1959 novel. It is a novel "indulging without restraint in the myth of the superweapon," while at the same time the mobile infantry of *Starship Troopers* is, because each soldier gets their own superweapon suit, fully "intoxicated by the myth of the omnipotent individual" when they suit up in their battle armor (2008, 52).

The combat suit worn by all of the soldiers in *The Forever War* is a direct science fiction descendent of that originally presented by Heinlein in *Starship Troopers*. It is not so much the suit that differs between the novels as it is the characters inhabiting them. Tim Blackmore described the difference:

> Haldeman's armour is identical to Heinlein's in that it is vastly multiplies the body's power. But Haldeman, a veteran foot soldier, won't forget how stupidly terrified people act, whether or not they're locked into death machines. (2005, 49)

But the suit in Heinlein's novel is, for all its imaginative similarity, still a far different artifact. Where Heinlein's suit is built with the contrivance that "you don't need to think about it," one of Mandella's trainers announces during his introduction to the technology makes the difference quite clear, noting the "fighting suit is the deadliest personal weapon ever built, and with no weapon is it easier to the user to kill himself through carelessness" (2004c, 16).

Thus, while Heinlein's narrative can offer a technological superweapon that allows its user to jump hundreds of feet into the air, assess an enemy while in motion, fire a shoulder-mounted battlefield nuclear weapon, all before landing, Haldeman's concerns are far more about the human element. While Heinlein's narrator can "walk, run, jump [and] lie down" the first time he wears his armor, Mandella's compatriots repeatedly die undertaking each activity. The German Americanist literature scholar Steffen Hantke has suggested that Haldeman "systematically debunks Heinlein's mythology of combat" with his emphasis on the perils of such weaponry (1998, 501). Indeed, Hankte's point, that "cadets die from mishandling the suit which is supposed to protect them, most of them in training rather than combat," makes the combat suits of Haldeman's fiction, for all their similarities, fundamentally different from those of Heinlein's. As a training officer tells Mandella and his fellow soldiers as they prepare to use their suits in near-absolute-zero temperatures:

> All you have to do is lean up against a boulder of frozen gas; there's lots of it around. The gas will sublime . . . and in about one-hundredth of a second, you have the equivalent of a hand grenade going off right below your neck. You'll never feel a thing. Variations on this theme have killed eleven people in the past two months.
>
> (Haldeman 2004c, 17)

Although such descriptions Haldeman refuses to allow his narrative to become the "glorious militarism" of Heinlein's and, in so doing, presents a narrative of war that is far more truthful than is found in *Starship Troopers*. Thus, Haldeman's conception of military technology avoids becoming a nearly dues ex machina superweapon. As Haldeman has noted elsewhere, "most of the things that are truly effective against the enemy tend to be effective against the user as well, if he trips or sneezes or forgets to put on the safety before he scratches his ear with the muzzle. (You think that doesn't happen?)" (Haldeman, Waugh, and Greenberg 1986, 4). And if Haldeman's version of the battle suit superweapon of science fiction is presented far more realistically than Heinlein's, his version of actual battles also derive almost directly from his experience in Vietnam. And it is with his presentation of armed conflict that *The Forever War* becomes, truly, a Vietnam novel.

For while *Forever War* does follow a similar narrative pattern to Heinlein's *Starship Troopers*, it diverges greatly thematically. Heinlein may have been almost enraptured by the technological implications of his battle suit and its use in promoting human Manifest Destiny across the frontiers of the galaxy, but Haldeman never uses such tropes in their traditional form. Instead, he uses those tropes repeatedly to explicitly undermine them with a sense of almost sarcastic irony. This shift in the vision of technology is one that is hardly surprising considering the Vietnam War. Philip Beidler, describing Norman Mailer's *Why Are We in Vietnam?* captures the sentiment not only of Mailer's novel but also of Haldeman's. Beidler sees Mailer's characters on an Alaskan hunt "that is at once a remembrance of a violent expansionist [American] past and a prophecy of the savage technowar soon to be fought in the jungles of Vietnam" (43–44). If there is a direct reaction in Haldeman to Heinlein's *Starship Troopers* it is simply the recognition of the morality of that "savage technowar" and where Heinlein delights in his battle suits, Haldeman recognizes them for what they are. Again, in Beidler's phrasing, Haldeman is aware that his characters, wearing their battle armor, are simply "armed with a storehouse of exquisite weaponry" that in reality does not confer moral superiority (44). Put another way, Haldeman is fully aware that the Vietnam War, from the perspective of the American literary response, is one that undermines the tropes of traditional American genres. Vietnam literature is, as John Clark Pratt described it, often the story of "American characters who enter Vietnam as traditional frontier huntsman, then become men trying merely to survive in a wilderness they do not understand" (1989, 238). In that, Beidler's "savage technowar" inherent in both Heinlein's and Haldeman's battle armor is merely a contrivance which may—or may not—aid survival. However, for Haldeman it is not the "glorious militarism" of Heinlein so much of the recognition that this was, again in Beidler's terms, "the biggest game there was: cowboys versus gooks, played in every moment for terrible stakes, and for reasons that no one would ever be able to explain" (1982, 15).

The American science fiction writer Connie Willis described the relationship between Heinlein's work and *The Forever War* by noting that "some critics see the book as a rebuttal to *Starship Troopers*, and I definitely think it is, but that's only one aspect of the book" (Willis 2006, xi). The difference for Willis is that *Forever War* is "a book filled with irony—because of the time jumps involved, a soldier could find himself obsolete during the course of a single war, or a single battle, and eternally separated from the things he was ostensibly fighting for" (Willis 2006, xi–xii). So while *Starship Troopers* is a text that Franklin can describe accurately as one in which "militarism animated every page; militarism—together with imperialism—is the novel's explicit message," *Forever War* is a novel that thematically undermines such a message by offering, also in a Franklin analysis, the "experience of becoming both extraterrestrial invaders of alien planets and exiles in time and space from planet Earth." (Franklin 1980, 112; 2000, 165). While *Starship Troopers* can be placed at the end of what Franklin calls Heinlein's period of "New Frontiers," *The Forever War* undermines not just the specifics of Heinlein's 1959 novel, but in offering a fantastic recounting of Haldeman's experiences in Vietnam it also undermines the idea of "New Frontiers" more generally. (For a description of Heinlein's "Few Frontiers" period see Franklin [1980, 64–125].)

This is hardly a surprising shift, however. As Richard Slotkin has pointed out, "the Vietnam War is a particularly appropriate symbol of the catastrophe that overtook the liberal consensus and the New Frontier" (1993, 636). Using standard science fiction themes built upon the extrapolation of endless "frontiers" as metaphors not for continuing opportunities, but for displacement and alienation is the method *The Forever War* uses to explain both the Vietnam War and its implications for American mythology. The fact that it is a science fiction novel, designed with what Willis called "a standard [science fiction] device—the relativistic effects of faster-than-light space travel—as a metaphor for the displacement and alienation of soldiers returning to a society with which they can no longer connect," might make it a correspondingly unrealistic depiction of Vietnam (xii). But simultaneously it also makes it an all the more truthful representation of that war's effect on the imperialistically driven ideology of frontiers that expand through the use of technology.

Haldeman places the beginnings of the narrative in the late 1990s, a time chosen "so that it would be barely possible that the officers and [noncommissioned officers] could be veterans of Vietnam" (Haldeman [2004a] unnumbered addition originally appearing in the 2001 British edition). Just before the events of the novel faster than light travel was developed on Earth, and in the cliché of science fiction, colonizing ships were sent into space. Almost immediately they encountered aliens that appeared hostile, and rather than following the cliché of the standard frontier and colonial adventure—what Mandela describes as "implementing a glorious adventure"—the characters of the novel instead must fight an interstellar war. (2004c, 8). Though thanks to works like Heinlein's *Starship Troopers* interstellar war is almost as clichéd a trope within science fiction as is interstellar colonization. Yet for Haldeman it will not retain anything approaching "glorious adventure" on any level. What follows is, as Franklin describes it, an extrapolation and fantasy of America's "longest war into . . . intergalactic combat instigated by generals and politicians, waged for profit, and conducted as a devastating fiasco from beginning to end" (2000, 165).

The use Haldeman makes of science fiction conventions for his own purposes must begin with his form of interstellar travel. From "hyperspace" to "warp drive" science fiction is replete with fictional ways around Einstein's constant allowing faster-than-light travel. *The Forever War* follows a quasi-scientific format of both instantaneous travel between "collapsars" and lengthy, relativistically time-slowing travel between those collapsar "jump points" in interstellar space. Each jump point in the novel is given a name of a letter from the Hebrew alphabet, allowing the narrator Mandella to take part in his first battle, for example, at a place designated "aleph-null." But, this nomenclature allows Haldeman to use the Hebrew letter "tet" with great frequency, emphasizing his Vietnam metaphor. Indeed, it is one of the latter battles that Mandela is involved in, Tet-2, which is the most destructive. In the later short story, "A Separate War," Haldeman has the character of MaryGay Potter describe the events briefly in a way that captures Haldeman's Vietnam metaphors:

> That was the Tet-2 campaign, which was a disaster, and William and I lay around doped to the gills with happyjuice while the others died their way through the disaster of Aleph-7. The score after the two battles was fifty-four dead, thirty-seven of us crips, two head cases, and only twelve more of less working soldiers, who were of course bristling with enthusiasm. (2005b, 125)

Throughout the course of Mandella's time as a soldier he spends the vast majority of his time traveling between such disastrous battles. This also allows Haldeman to emphasize sociological change on Earth during the war as a metaphor for similar changes in the United States while Haldeman was in Vietnam. At one point Mandella comments, "I'd been in the army ten years, though it felt like less than two. Time dilation, of course; even with the collapsar jumps, travelling from star to star eats up the calendar" (2004c, 80). Thus, while drafted in 1996, and experiencing roughly two years subjectively in the army, Mandella could find the external world has moved through ten years and, suddenly, although he remains relatively young it is 2007 on Earth. Not surprisingly, this is a major source of feelings of alienation for the character. One study of Vietnam Veterans described this kind of alienation from the process of fighting the war in stark numbers, reporting a tendency to feel "somewhat (51%) or very (34%) isolated when they returned home" (Frey-Wouters and Laufer 1986, 41). Haldeman accomplishes this sense of isolation through the science fiction convention of "time dilation." By the time Mandella returns to Earth it is well into the early twenty-first century, his father is dead, and there has been a "Ration War" due to crop failures and overpopulation similar to the predictions of Paul Ehrlich that will be discussed in chapter six. Furthermore, the various governments have begun promoting homosexuality as a method of population control, and his mother is living with her same-gender lover. Mandella considers what an army captain tells him of a life on Earth where, "about a third of everybody in Europe and America" is now gay:

> He said the relations between people who chose homolife and the ones he called "breeders" were quite smooth, but I wondered. I never had much trouble accepting homosexuals myself, but then I'd never had to cope with such an abundance of them.
>
> (Haldeman 2004c, 111, 112, 114)[4]

The prevalent homosexuality Mandella finds on his return to Earth is meant specifically as a form of alienation, but it must also be read as a statement about the way in which the western genre has been used in science fiction. Melody Graulich has claimed, the entire "myth of the frontier dream has been formulated from the male point of view," and as a result that point of view has invariably been a hyper-masculinized heterosexuality (1989, 187). Within the confines of strict genre norms Roger Horrocks has been more overt, calling "the western novel and film . . . 'phallic discourses' taken to an endpoint—men gaze at each other, pump bullets into each other's bodies, and lust after women as bar-room 'whores'" (1995, 56). While Horrocks quickly notes that there remains an unsubtle homoeroticism in such narratives, Haldeman takes any subtle convention that might remain and removes it. By making it obvious that Mandella's alienation is the partial result of homosexuality—and that the homosexuality in question is both a positive development and a morally reasonable one—Haldeman has undermined yet another of the main science fiction tropes derived from the western novel and film. Such a narrative, traditionally within the genre of the western or in science fiction, is about what John Cawelti called the "effort involved in reaffirming a masculinity that seems continually besieged by temptations usually coded in modern American popular culture as feminine" (1999, 159). The tropes of both the western and earlier science fiction would have Mandella use his "skill in violence" and then choose "to act in such a way as to further a social order based on the ideals of progress and success" or else ride "off into the desert" (ibid., 53) In either case a masculine heterosexuality would be what such a hero was either advancing or achieving. When Mandella returns to Earth from the war—like many Vietnam veterans including Haldeman himself—he finds the new sexual mores that confront him to be, literally, anathema to those he thought he was explicitly fighting for. The abundant homosexuality of the novel is, then, as much a statement about the alienation of returning veterans, as it is an even more blatant evisceration of the gender paradigms of the western-derived science-fiction frontier trope.

Yet this will not be the only method in which Haldeman offers his fantastic version of the Vietnam War to undermine such frontier tropes. While experiencing the alienation of such new sexual conventions Mandella spends just a few short weeks on Earth as his mother dies of treatable illness due to a lack of medical attention because her age and lack of worth in the war effort leaves her uninsured. During that time he is forced to buy a gun for personal protection, and it is not just any gun, either, but a "dinosaur stopper" with "authentic Old West styling" (Haldeman 2004c, 121). No longer what John Cawelti called the "symbolic expression of masculine potency in an unmistakable way," Mandella's gun is one that cannot be used, as Cawelti described it in the fictive terms of American Old West fiction, "with precise and powerful effectiveness" (Cawelti 1999, 39; Cawelti 1984, 89). He uses it to disrupt a street rape—"I remember trying to aim for his arm"—and the result is, monstrously, "the blast hit his shoulder and ripped off his arm and what seemed to be half of his chest" (2004c, 134).

That his "authentic Old West"–styled gun cannot offer the mythical resonance of such a weapon is made plainer when, later, Mandella journeys to what can only be read as the frontier land of South Dakota—a location, Haldeman informed me, selected because it is "just one of them big square states out west"—to live with MaryGay and her parents on their communal farm (2011b). While there he works on the family homestead in almost traditional western fiction fashion, plowing under five acres of farmland. Although it is "hard, hardening work, but pleasant" this version of the frontier lasts little longer than the one implied by Mandella's purchase of a gun (2004c, 143). "Jumpers," described as criminals who "take advantage of our relative isolation" and attack such farming homesteads, attack the farm (2004c, 142). They cannot be read as other than the American Frontier concept of mining "claim jumpers." MaryGay's mother and father are both killed when the jumpers attack their communal farm, and another version of the frontier is lost to both MaryGay and Mandella.

For Mandella Earth cannot be a place of frontier opportunity, or even frontier justice with his dinosaur stopper gun. But after the death of his mother he talks to his brother on the moon who describes it within a traditional frontier framework: "Come to Luna," his brother suggests. Where Earth has lost any remnant of the frontier mentality inherent in so much of America science fiction, the moon is a place, according to Mandella's brother, where "you can still be an individual. Where we don't throw people out the airlock on their seventieth birthday" (ibid, 149–150). Indeed, it is another common trope of American science fiction that the moon is presented as another new frontier, and Haldeman offers the suggestion of just such a paradigm. Thus, William Mandella and MaryGay Potter reenlist as combat trainers with a posting on the moon. That posting lasts only until their arrival on the moon where they are immediately reassigned to combat duty. The moon, it seems, is not the frontier that science fiction would hope, or the one that NASA will plan for in the 1960s as described in chapter four. For Haldeman it is, simply, another science fiction cliché that can be used to both further his narrative, and undermines the traditional tropes of the genre.

If Mandella's time on Earth, with the alienation felt in an emerging homosexual culture, the lack of anything remotely resembling frontier opportunity, and the deaths of both his and MaryGay's family members serves to highlight the estrangement felt by returning Vietnam veterans, his time in active duty combat highlights the experience of many soldiers in Vietnam, itself. Over the course of the narrative Mandella takes part in three battles. It is important to indicate, merely, "takes part." He fights, yes, but reluctantly and haphazardly except when posthypnotic training has been used to accentuate his killing instinct. And many of these battles mirror specific battles of Vietnam or images of the war generally.

MY LAI AND TET IN THE FRONTIER OF SPACE

The first time Mandella sees something akin to fighting Haldeman makes clear his larger theme, for it is not combat so much as a massacre. The alien Taurans do not fight the good fight of "glorious militarism" but are simply killed. Franklin draws the comparison simply, noting that the humans

> butcher every single Tauran they encounter in the first actual combat contact between the two species, mirroring the U.S. slaughter at My Lai and countless other Vietnamese villages. Most of the deaths suffered by the soldiers from Earth result from their traumatic revulsion against what the military and their government have made them do. (2000, 165)

Having been fed a posthypnotic training regimen Mandella and his fellow soldiers have no difficulty committing an atrocity against aliens that, for the most part, seem incapable of fighting back. Indeed, as Haldeman presents them, the alien enemies at first seem almost unaware that they are considered exactly that: enemies. "They milled around," Mandella offers, "about a hundred of them, and watched us get closer" (2004c, 68). Attempting to capture one alien for "interrogation," Mandella and his compatriots work hard at isolating a target, although that plan quickly devolves into chaos. "I hated myself for smiling. It was a slaughter," Mandella notes at one point, "even though our flank was outnumbered five to one" (ibid, 69). Like the events at My Lai, where victims were herded together in rice paddies or ditches to be shot, and where soldiers would use grenade launchers to kill Vietnamese running away, Mandella and his fellow soldiers follow a similar pattern. The testimony of soldier Dennis Conti at the My Lai court-martial of Lieutenant William Calley could be actual text from Haldeman's scene: "Lots of heads was shot off, pieces of heads and pieces of flesh flew off the sides and arms," Condi testified. "Then I saw a group of five women and six kids—eleven in all—going to a tree line. 'Get 'em! Get 'em! Kill 'em!' Calley told me. I waited until they got to the line and fired off four or five grenades" (Linder, n.d.).[5] Haldeman has Mandella describe his scene similarly:

> [B]odies and parts of bodies [were] piled up high . . . My helmet reverberated with hysterical laughter while we slashed them to gory chunks . . . "We've got our prisoner," Cortez shouted. "*Kill*!" . . . One fell, sliced in two, but the others, about ten of them, kept going and were almost to the doors when the grenadiers started firing. (2004c, 69–70)

Later, after the soldiers have been brought out of their posthypnotic state that enabled them to undertake the massacre Mandella explains why so many "almost went crazy with the memories of bloody murder multiplied a hundred times":

> Because it *was* murder, unadorned butchery—once we had the anti-spacecraft weapon doped out, we hadn't been in any danger. The Taurans hadn't seemed to have any conception of person-to-person fighting. We had just herded them up and slaughtered them.
>
> (ibid., 72; emphasis in the original)

If Philip Beidler is correct that some events during the Vietnam War were so "monstrously concrete and immediate that they could only be handled by imaginative conversion into some unreal other," then it would seem that Joe Haldeman has offered exactly what Beidler believes necessary. As Franklin has put it, "Haldeman explodes the pet practices and illusions of U.S. militarism by taking them to absurdly fantastic dimensions" (2000, 165). And Haldeman's conceit of a posthypnotic treatment forcing his narrator's actions does not provide a useful unreal excuse in this "conversion into some unreal other," either. Mandella is still wracked with guilt over his actions, allowing Haldeman to make his connection between his scene of slaughter and those from the real world he is referencing. "Back in the twentieth century, they had established to everybody's satisfaction that 'I was just following orders' was an inadequate excuse for inhuman conduct," Mandella notes as he discounts the science fiction convention of his officers having compelled his behavior through hypnosis and moving quickly to massacres that he can assume his readers familiarity with such as My Lai. "Worst of all was the feeling that perhaps my actions weren't all that inhuman. Ancestors only a few generations back would have done the same thing, even to their fellow men, without any hypnotic conditioning" (Haldeman 2004c, 73).

And perhaps Mandella's final statement after taking part in a massacre also nicely sums up a more general reaction to the American involvement in Vietnam: "I was disgusted with the human race, disgusted with the army and horrified at the prospect of living with myself for another century or so" (ibid., 73).

It is wholly unsurprising, considering the internal conventions of the novel that eventually the now lieutenant Mandella would fight in a battle called, simply, "Tet-2." It is even less surprising, considering Haldeman landed in Vietnam just as the original Tet campaign was ending, that the battle would be, as MaryGay would call it, "a disaster." Haldeman offered the entirety of his Tet-2 campaign roughly two pages of text. Mandella is landing on an unnamed alien planet, there is a countdown from five which never passes three, then "we were rolling, horribly, pieces of people and ship scattering . . . and I was rolling free, short stump of a leg trailing blood that froze shiny black on the dull black rock" (ibid., 158). The humans do survive the battle, and Mandella and MaryGay are evacuated. In Heinlein's *Starship Troopers* the planet where soldiers are offered rest, recuperation, and relaxation is called, simply, "Sanctuary." In *The Forever War* it is called "Heaven," and Haldeman describes it in detail, and in ways strikingly like the "virgin land" once used to describe the New World. "Heaven was a lovely, unspoiled Earth-like world," Mandella proclaims. It is a world, he continues, with "Virgin forests, white beaches, pristine deserts. . . . We landed, as everyone does, at the jungle city, Threshold . . . The only sign of civilization was a short runway that suddenly appeared" (ibid., 161).

In the hospital city of Threshold, Mandella has his leg slowly regrown while MaryGay has a similar procedure for the arm she also lost during Tet-2. Then they are allowed to explore the planet of Heaven—a planet, it seems, specifically designed by Haldeman to present the best possible image of the frontier. "We had fun, desperate fun," Mandella offers.

> We rented a flyer and camping gear and went off for weeks, exploring the planet. There were icy rivers to swim and lush jungles to crawl through; meadows and mountains and polar wastes and deserts. (ibid., 167)

Not merely a frontier, however, the planet Heaven is a frontier designed for human enjoyment. In ways that will be echoed by Gerard K. O'Neill's "High Frontier" described in chapter six, the planet is maintained, "with all the dangerous and uncomfortable elements eliminated." Thus, large "pressor fields" "kept out large predators and such insect life that was not necessary for the health of the plants inside" the city of Threshold (ibid., 161–162). And when exploring Heaven Mandella and MaryGay could use "individual pressor fields" allowing them to "sleep naked in a blizzard" or, at their whim, "they could just take nature straight" (ibid., 167).

After "taking nature straight" on Heaven, their injuries fixed, both Mandella and MaryGay are assigned to yet further combat. Mandella is promoted to the rank of major, separated from MaryGay, and sent on the longest range mission in the entirety of the Forever War. As he prepares he is told he has "survived two-and-a-half more" battles than the average soldier, explaining his promotion. The officer assigned to explain the historical changes since Mandella was drafted informs him that his survival will likely make him, through the work of the army's "propaganda boys" a "folk hero." Ironically, Mandella replies with a response designed to accentuate that this is, indeed, not the traditional frontier of American science fiction: " 'Folk hero.' I sipped my beer. 'Where is John Wayne now that we really need him?' " (ibid., 179). In charge of more than one hundred soldiers he is sent to the "enemy's frontier," literally outside of the Milky Way galaxy. While there, attempting to hold territory against enemy incursion, Mandella considers his poor tactical situation by thinking through his training in fully ironic fashion. That training leads him to think, "Keep your head down and your chin up and pray for the cavalry." That the cavalry might not arrive is, obviously, a major concern, but as Mandella notes, he and his troops are required to "hold your position," and—again, fully ironically—"don't contemplate Dien Bien Phu [or] the Alamo" (ibid, 228; in the published version Haldeman also includes the Battle of Hastings, and offers "Dien Bien Phu" in the older format of a single word. It has been changed to the contemporary format here).

Offering both the notorious 1954 French defeat in their loss of their colonial holding of Vietnam and the Texan defeat against the Mexican army in 1836 together, Haldeman has nicely combined both his imagery of the Vietnam War and that of the traditional frontier narrative. Confronted with a battle he fully expects to lose—and, as a result, die in—Mandella cannot help but consider similar hopeless battles. Of course, he had already been informed that the winning of wars is not exclusively about the winning of such battles. Another officer, describing the Forever War—and offering a precise commentary on Vietnam in the process—had told Mandella that wars "are won not by a simple series of battles won, but by a complex interrelationship among military victory, economic pressures, logistic maneuvering, access to the enemy's information, political postures—dozens, literally dozens of factors" (2004c, 100). So, on a planet at the farthest edge of the frontier for both humans and the alien Taurans, Mandella has to confront the idea that he fully expects to die in a battle of little or no consequence.

Where Heinlein might offer a "glorious militarism" in *Starship Troopers*, then, Haldeman offers a far more sanguine version of technowar. Heinlein is representative of an entire trope of science fiction. It is one that imagines the use of technology to master another frontier, often through science fictional wars of imperialism. That trope centers on use of military technology to begin the process of the "conquest" of space, or what Mogen described as the theme of enlisting "the energies of our frontier heritage to conquer the new wilderness of outer space that technology has opened to us." Haldeman fully undermines this theme as well. Offering his soldiers battle armor similar to Heinlein's, a war against an alien enemy, and science fiction weapons of fantastic ability, Haldeman places the final battle of his novel in a place where none of this technology is of significant use at all. Haldeman provides another science fiction convention, a "stasis field" that neutralizes all electrical and magnetic activity, including those existing within a living body. As a result the complex battle armor becomes little more than a shielding device allowing the soldiers to continue to live. Explosives, guns, lasers and all other imaginable weapons are useless—there can be no more technowar at all. As a result the human soldiers and their alien enemies are left fighting each other without the fetishistically described weapons usually found in science fiction. "We'd killed more than half of them with arrows and spears," Mandella notes at one point. "I drew my sword and waited" (ibid., 239).

No longer, then, is *The Forever War* in any way keeping pace with the clichés of American frontier science fiction. By making it a novel about the Vietnam War Haldeman could not keep the tropes of such fiction in place and make his statement come across in any way truthfully. The technology of such fiction—and the technowar of the Vietnam experience—is obliterated by the needs of the story. The imagery of the endless frontier of space, and wars to conquer that frontier are also lost like similar themes that were destroyed in the jungles and rice paddies of Vietnam. In the final analysis for Haldeman *The Forever War* is "about Vietnam because that's the war the author was in." But at the same time he notes, "it's mainly about war, about soldiers, and about the reasons we think we need them" (2004a). And on that level Haldeman has often made it clear that there is little difference between "a boy returned rotting on his shield or hidden antiseptically inside a plastic bag" (1979, 1). In having his protagonist, Mandella, finish his final battle with a sword in his had he makes that point most clearly.

The one remaining trope of such American science fiction is that the expanding frontiers are supposed to offer new planets to colonies. This is what Mogen, writing about another science fiction author but in a way applicable to many within the genre, calls "visionary fragments about flowering New Worlds . . . combined with technological nuts and bolts, so that the center of the rhetorical machinery is the old image of the New World" (1993, 37). To offer such images, however, Haldeman's Vietnam novel would have to imply that the war in Vietnam could have offered something similar. Here again he fully reverses the trope to his own ends and once again undermines the notion of the frontier not just in American science fiction, but in American culture more generally.

In the tradition novel of American science fiction the narrative would end either at the height of glorious battle—as does Heinlein's *Starship Troopers*—or with the end of the conflict and the acquisition of new worlds to colonies for an expanding human Manifest Destiny. In either case the knowledge of that Manifest Destiny would still be a major influence within the text and the characters—and the readers—would be aware of an expanding human civilization. Haldeman's experience in Vietnam means neither option is available in *The Forever War*. In an ironic way, however, Haldeman deals with all of the tropes inherent in such texts. Mandella and MaryGay do end up, together, after the war. They do end up on a new planet that a bartender describes to Mandella as "A very nice place. Garden planet."[6] But the human civilization that they were supposedly fighting for is now something to which they can never belong. Thus, Mandella's "Garden planet" is, as MaryGay describes it to him in a letter inviting him to join her there, "a kind of Coventry for heterosexuals" and a "genetic control baseline" (Haldeman 2004c, 252). They will be allowed to live out their lives there not as part of an expanding human civilization but as the holders of "backup" genes should the broader human species require their diversity. In a final statement about soldiers in war, and the Vietnam War itself, Haldeman finally rejects all of the tropes he has adopted from traditional science fiction. There is no "glorious militarism," there is no "new wilderness of outer space that technology has opened to us," as Mogen put it. So the planet on which Mandella and MaryGay retire from their time in the Forever War is named to provide Haldeman's final statement about Vietnam. And, at the same time, it can be read as a statement about the status of the frontier myth after Vietnam.

It is a planet called "Middle Finger."

Within the novel's narrative Haldeman is "flipping off" his experience in Vietnam with such a name, as well as the experience of war in general. Yet, in using the conventions of science fiction that were built on the myth of the frontier, he is doing something else as well. He is also "flipping off" the mythological status of what seemed to drive the United States to war in Vietnam, that is, the frontier ideology that inspired it. Haldeman's desire—or, as both Franklin and Blackmore describe it, his "delight"—in subverting such established tropes and using them, essentially, as an attack on the myths that inspired them is certainly a statement about the place those myths seem to have as the war in Vietnam came to a close. While other Vietnam veterans were busy using the accepted forms of the American literary tradition in an attempt to explain Vietnam, and American scholars of literature would soon be busy categorizing those works within the American canon, Haldeman was offering quite the opposite. Scholars of American mythology should have little trouble with such a statement about those myths—and indeed, they did not.

Myths, however, are far hardier cultural manifestations than the reports of the demise of "frontier myths" would suggest. They often change but rarely die. So while scholars were busy burying the traditions of the "frontier"—whether by ending the Myth and Symbol school of American Studies or, once again, dissecting the "frontier thesis" of Frederick Jackson Turner—much broader parts of American society were not so much dismissing the frontier as altering it, adjusting it, and adapting it for new situations. Events like Tet and My Lai—and the Vietnam War as a whole—did not easily fit within established mythological structures. The *Time-Life* reporter Hugh Sidey, reporting on My Lai in 1969 for *Life* magazine, offered a statement which he attributed to then–national security advisor Henry Kissinger. My Lai, Sidey reported, was simply "not within the [American] national experience" (1969, 4). Myths remain purposeful only when new experiences can be fitted into their explanatory framework. William H. McNeill had worried about the fate of such myths—and the way scholars had reacted to Vietnam by ignoring them—in his "Care and Repair" essay. In it he nicely captured the intellectual response to Vietnam by noting that "the effect of such scholarship is to substitute a divisive for a unifying myth" (1986, 33). McNeill was primarily worried about professional historians' disdain for both "myth" and "general statements about the world" that myths provide. Haldeman's *Forever War* offered a novelistic statement about such myths as well, and it is one that agrees with McNeill's worries. Vietnam, for both McNeill and Haldeman, seemed to be an exercise in "discrediting old myths without finding new ones to replace them" (ibid., 25). McNeill's essay, however, was originally published in *Foreign Affairs* in 1982. He was worried, as he put it, that "inherited ideas—whether dating back to pagan Greece, Christian Europe, 1776 or 1948—are simply inadequate, and there is no use pretending otherwise" in the wake of Vietnam (1986, 42). And, in *The Forever War* Haldeman offers much the same conclusion. American science fiction, which specialized in Doty's "forms of the future," was, Haldeman indicated, no longer capable of providing those myths in the traditional form. Vietnam had undermined the traditional mythology, and while Haldeman offers none of Marx's "new symbols of possibility" in *The Forever War*, he does firmly say the "old symbols" are no longer working. But American society, as a whole and as subsets of that whole, was not in the process of fully rejecting those myths. It was in the process of experimenting with new forms of them.

While Haldeman was serving in Vietnam NASA would be planning their missions that would lead to the Apollo 11 landing shortly after Haldeman's return. *Star Trek* would be airing its second and the beginning of its third seasons. And Gerard K. O'Neill would be trying, hard, to become part of NASA's mission, while watching *Star Trek*, and planning a High Frontier all his own.

To academics in history, American Studies, and other fields it did appear as though American frontier myths had been shattered by Vietnam. But other facets of American culture were by no means ready to let it be simply thrown in the dustbin of history. While scholars and members of the intelligentsia were more than willing to jettison any notion of such myths as flawed and dangerous many cultures in America were more than willing to simply revise those myths to new ends—which remains a standard method of the continuing utilization of mythic forms. If Vietnam seemed to fully obliterate those frontier myths for many—and, for a time, Vietnam seemed to do exactly that—the cultures of the United States quickly offered alternatives. These alternatives were not new myths, but a closely related series of myths with a very similar underlying structure based, still, on ideas of the "frontier." The cul-de-sac of American myth envisioned by scholars quickly offered a new route to possible new frontiers over even further horizons. First, NASA would offer a very traditional version of a frontier myth with the traditional riches such frontiers were supposed to offer. *Star Trek* followed this with its more refined version that rejected the tradition of the frontier offering a new version of paradise. And a few short years later Gerard K. O'Neill would present a whole new garden in the machine and a frontier he believed capable of solving all the world's ills. It is in the exploration of those alternative myths that the rest of this work is committed to examining.

NOTES

1 Specific statistics vary tremendously by report, especially in terms of casualty numbers of Viet Cong and the North Vietnamese Armed Forces. It is an interesting point that many American undergraduate textbooks on the Vietnam War go to great lengths to offer specific statistics of American combat casualties but rarely attempt to offer similar exactitude for those of the North Vietnamese.
2 Like the Tet Offensive, the My Lai massacre has resulted in a significant literature. When not specified otherwise details are drawn from the following: Oliver (2006) and Bilton and Sim (1993); issues of moral and ethical scholarship are examined in Oliver (2004).
3 Hellmann (1986, 92).
4 Haldeman (2004c, 111, 112, 114).
5 University of Missouri, Kansas City, Professor of Law Doug Linder maintains a web resource for his students, from which this quote is taken. See Linder (n.d.); I confirmed the quote, however, with one of Professor Linder's colleagues at the same university, William G. Eckhardt. Professor Eckhardt was also prosecutor at the My Lai courts-martial (Eckhardt 2011).
6 Haldeman (2004c, 253).

WORKS CITED

Abbott, Carl. 2006. *Frontiers Past and Future: Science Fiction and the American West*. Lawrence: University Press of Kansas.
Anderegg, Michael A. 1991. *Inventing Vietnam: The War in Film and Television*. Philadelphia: Temple University Press.
Anderson, Dale. 2006. *The Tet Offensive: Turning Point of the Vietnam War*. Mankato, MN: Compass Point Books.
Beidler, Philip D. 1982. *American Literature and the Experience of Vietnam*. Athens: University of Georgia Press.
———. 1991. *Re-Writing America: Vietnam Authors in Their Generation*. Athens: University of Georgia Press.
Bilton, Michael, and Kevin Sim. 1993. *Four Hours in My Lai*. New York: Penguin.
Blackmore, Tim. 1993. "Warring Stories: Fighting for Truth in the Science Fiction of Joe Haldeman." *Extrapolation* 34 (2): 131–46.

———. 2005. *War X: Human Extensions in Battlespace.* Toronto: University of Toronto Press.

Broyles, William. 1996. *Brothers in Arms: A Journey from War to Peace.* Austin: University of Texas Press.

Cawelti, John G. 1984. *The Six-Gun Mystique.* Bowling Green: Bowling Green State University Popular Press.

———. 1999. *The Six-Gun Mystique Sequel.* Bowling Green: Bowling Green State University Popular Press.

Creek, Mardena Bridges. 1982. "Myth, Wound, Accommodation : American Literary Response to the War in Vietnam (Ph.D. Dissertation)." Muncie, Indiana: Ball State University.

Dann, Jack, and Jeanne Van Buren Dann. 1987. "Introduction." In *In the Field of Fire*, edited by Jeanne Van Buren Dann and Jack Dann, 9–16. New York: Tor.

Devine, Jeremy M. 1999. *Vietnam at 24 Frames a Second: A Critical and Thematic Analysis of over 400 Films about the Vietnam War.* Austin: University of Texas Press.

Eckhardt, William G. 2011. "My Lai Trail Transcripts, Email to Matthew Wilhelm Kapell," October 13.

FitzGerald, Frances. 1972. *Fire in the Lake: The Vietnamese and the Americans in Vietnam.* New York: Vintage.

Franklin, H. Bruce. 1980. *Robert A. Heinlein: America as Science Fiction.* New York: Oxford University Press.

———. 1993. *M.I.A., Or, Mythmaking in America.* Expanded and updated ed. New Brunswick, NJ: Rutgers University Press.

———. 1995. "Introduction." In *Future Perfect: American Science Fiction of the Nineteenth Century: An Anthology*, Revised and Expanded edition, 1–4. New Brunswick, NJ: Rutgers University Press.

———. 2000. *Vietnam and Other American Fantasies.* Amherst: University of Massachusetts Press.

———. 2007. " 'Vietnam' in the New American Century." In *The United States and the Legacy of the Vietnam War*, edited by Jon Roper, 33–50. New York: Palgrave Macmillan.

———. 2008. *War Stars: The Superweapon and the American Imagination.* Revised and Expanded. Amherst: University of Massachusetts Press.

Frey-Wouters, Ellen, and Robert S. Laufer. 1986. *Legacy of a War: The American Soldier in Vietnam.* Armonk, NY: M.E. Sharpe.

Gordon, Joan. 1980. *Joe Haldeman.* Rockville, MD: Wildside Press.

Graulich, Melody. 1989. " 'O Beautiful for Spacious Guys': An Essay on the 'Legitimate Inclinations of the Sexes.' " In *The Frontier Experience and the American Dream*, edited by David Mogen, Mark Busby, and Paul Bryant, 186–201. College Station: Texas A&M University Press.

Haldeman, Joe. 1979. "Introduction." In *Study War No More.* London: Futura.

———. 1993. "Vietnam and Other Alien Worlds." In *Vietnam and Other Alien Worlds.* Framingham, MA: NESFA Press.

———. 2004a. "Author's Note, British Gollanz 'SF Masterworks' Edition." In *The Forever War* (unnumbered pages). London: Gollancz.

———. 2004b. "Interim Report: An Autobiographical Sketch." In *Gale Contemporary Authors Series.* Farmington Hills, MI: Gale. http://home.earthlink.net/~haldeman/biolong.html.

———. 2004c. *The Forever War*. London: Gollancz.

———. 2005a. "1968: A Novel." In *War Stories*, 173–398. San Francisco: Nightshade Books.

———. 2005b. "A Separate War." In *War Stories*, 125–54. San Francisco: Nightshade Books.

———. 2005c. "DX (poem)." In *War Stories*, 163–70. San Francisco: Nightshade Books.

———. 2005d. "Introduction: A Tangled Web." In *War Stories*, 171–72. San Francisco: Nightshade Books.

———. 2005e. "Introduction: You Have To Start Somewhere." In *War Stories*, 1–4. San Francisco: Nightshade Books.

———. 2005f. "War Year." In *War Stories*, 5–72. San Francisco: Nightshade Books.

———. 2011a. "Email on Star Trek, Vietnam, and The Forever War, to Matthew Wilhelm Kapell," May 6.

———. 2011b. "Email on The Forever War Locations and Starship Troopers, To Matthew Wilhelm Kapell," October 11.

Haldeman, Joe, Charles G. Waugh, and Martin Harry Greenberg, eds. 1986. *Body Armor/2000*. New York: Ace Science Fiction Books.

Hallin, Daniel C. 1986. *The "Uncensored War": The Media and Vietnam*. New York: Oxford University Press.

Hantke, Steffen H. 1998. "Surgical Strakes and Prosthetic Warriors: The Soldier's Body in Contemporary Science Fiction." *Science Fiction Studies* 25 (3): 495–509.

Heinlein, Robert A. 2005. *Starship Troopers*. London: Hodder and Stoughton Ltd.

Hellmann, John. 1986. *American Myth and the Legacy of Vietnam*. New York: Columbia University Press.

Herzog, Tobey C. 1992. *Vietnam War Stories: Innocence Lost*. London: Routledge.

Horrocks, Roger. 1995. *Male Myths and Icons: Masculinity in Popular Culture*. New York: Palgrave Macmillan.

Katerberg, William H. 2008. *Future West: Utopia and Apocalypse in Frontier Science Fiction*. Lawrence: University Press of Kansas.

Kellogg, Ray, and John Wayne. 1968. *The Green Berets*. Burbank, CA: Warner Bros.

Kilgore, De Witt Douglas. 2003. *Astrofuturism: Science, Race, and Visions of Utopia in Space*. Philadelphia: University of Pennsylvania Press.

Klinkowitz, Jerome. 1980. *The American 1960's: Imaginative Acts in a Decade of Change*. Ames: Iowa State University Press.

Lind, Michael. 1999. *Vietnam, the Necessary War: A Reinterpretation of America's Most Disastrous Military Conflict*. New York: Free Press.

Linder, Doug. n.d. "Dennis Conti, Witness for the Prosecution." *Famous Trials Web Resource, University of Missouri, Kansas City, School of Law*. http://law2.umkc.edu/faculty/projects/ftrials/mylai/Myl_tConT.htm.

Matusow, Allen J. 1984. *The Unraveling of America: A History of Liberalism in the 1960s*. New York: Harper & Row.

McNeill, William H. 1986. "The Care and Repair of Public Myth." In *Mythistory and Other Essays*, 23–42. Chicago: University of Chicago Press.

Melling, Philip H. 1990. *Vietnam in American Literature*. Boston: Twayne Publishers.

Mogen, David. 1993. *Wilderness Visions: The Western Theme in Science Fiction Literature*. 2nd ed., revised and expanded. Rockville, MD: Wildside Press.

Myers, Thomas. 1988. *Walking Point: American Narratives of Vietnam*. New York: Oxford University Press.

Oliver, Kendrick. 2004. "Towards a New Moral History of the Vietnam War?" *The Historical Journal* 47 (3): 757–74.

———. 2006. *The My Lai Massacre in American History and Memory*. Manchester: Manchester University Press.

Panshin, Alexei. 1968. *Heinlein in Dimension: A Critical Analysis*. 1st ed. Chicago: Advent Publishers.

Pratt, John Clark. 1989. "The Lost Frontier: American Myth in the Literature of Vietnam." In *The Frontier Experience and the American Dream*, edited by David Mogen, Mark Busby, and Paul Bryant, 236–427. College Station: Texas A&M University Press.

Roper, Jon. 2007. "Over Thirty Years." In *The United States and the Legacy of the Vietnam War*, edited by Jon Roper, 1–20. New York: Palgrave Macmillan.

Schmitz, David F. 2005. *The Tet Offensive: Politics, War, and Public Opinion*. New York: Rowman & Littlefield.

Schorer, Mark. 1968. "Appendix: The Necessity of Myth." In *Myth and Mythmaking*, edited by Henry Alexander Murray, 354–58. Boston: Beacon Press.

Schweitzer, Darrell. 1977. "An Interview with Joe Haldeman." *Science Fiction Review* 20 (February): 26–30.

Sidey, Hugh. 1969. "The Presidency: In The Shadow of Mylai." *Life*, December 12.

Slotkin, Richard. 1993. *Gunfighter Nation: The Myth of the Frontier in Twentieth-Century America*. New York: HarperPerennial.

Willis, Connie. 2006. "Meet Joe Haldeman." In *A Separate War & Other Stories*, xi–xvi. New York: Ace.

Wilson, James C. 1982. *Vietnam in Prose and Film*. Jefferson: McFarland.

Wingo, Hal. 1969. "The Massacre at My Lai." *Life*, December 5.

4 Technological Triumph, Mythological Miasma

NASA, the Moon, and Transforming Mythos into Logos

> But it was a different sort of beauty than I had ever before seen, *Magnificent*, I thought, then said, "Magnificent desolation." It was a spontaneous utterance, an oxymoron that would take on ever-deeper dimensions of meaning in describing this strange new environment.
>
> —Buzz Aldrin (2009, 34)

Time-Life reporter Hugh Sidey just wanted an answer to a question.

While it would be an exaggeration to suggest that the question—and its eventual answer—make Sidey one of the more important people to propel the United States space effort in the 1960s, it also would not be far off the mark to say that, in probing for his answer, Sidey pushed President John F. Kennedy to reach for the moon. Originally Sidey was a beat reporter for *Life* magazine, but by the early 1960s he had become the White House correspondent for its sister publication *Time*, and on April 14, 1961, he was the reporter charged with covering the original seven Mercury astronauts. Invited to the White House just two days after Soviet cosmonaut Yuri Gagarin shocked Americans and the world by spending 108 minutes in a single orbit of the Earth aboard his Vostok capsule, Sidey was allowed to sit in on meetings both with Kennedy's top advisors, and then with Kennedy himself, as plans were discussed on how to "catch up" to the Soviet accomplishment (Logsdon 2010, 75–76).[1]

With Sidey in those meetings were Kennedy's science advisor Jerome Wiesner, Presidential Special Counsel Ted Sorensen, and the National Aeronautics and Space Administration's head James E. Webb, among others. Sidey also asked for an interview with Kennedy and, as part of that process, prepared a memo for the president's press secretary, Pierre Salinger. Sidey presented Salinger with a number of questions, all of which seem perfectly reasonable from a reporter trying to find a story to report on. But, from another perspective, the questions Sidey asked also could be interpreted as a certain kind of goading. He noted that "a lot of good minds in NASA . . . think we are still fiddling, haven't made the necessary decisions . . . They claim, with compelling logic, that if we are to get in this race at all we've got to declare a national space goal." Sidey emphasized the national worry about the Gagarin flight central to his questions in that memo and suggested a time frame that fell within the decade. He pointed out that without a "national space goal" the United States was in a position where "we are going to sit here over the next eight years and watch the Soviets march right on ahead." (Logsdon 2010, 75–76).[2] In short, Sidey was asking, "What are we going to do in response to the Gagarin flight?" Jerome Wiesner would draft two memos to the president about meetings that day, the second as a response to Sidey's questions, which affirmed a need for "establishing clear-cut national goals to go beyond present plans" (Logsdon 2010, 76).

Hugh Sidey would report about those meetings, first in *Life* magazine and, later, in his own book on Kennedy. At the end of the second meeting of the day, the one with both Kennedy and Sidey present, Kennedy is reported to have said, "[W]hen we know more, I can decide whether it is worth it or not. If somebody can just tell me how to catch up. Let's find somebody—anybody. I don't care if it's the janitor over there" (Sidey 1961, 20; 1963, 102). Then, Sidey reported, the president left the meeting with his advisor Ted Sorensen, and both went to what was then usually called the Oval Room. A few minutes later Sorensen left the Oval, found Hugh Sidey waiting for him, and Sidey asked, again, for an answer to his original question. For all intents and purposes, all Sidey seemed to want was a story to file with his editors—that is, to do the job of a reporter. But, in repeatedly asking the question Sidey got his answer from Sorenson, and through Sorensen from President Kennedy. As a result Hugh Sidey holds a place as one of the people who pushed for an answer and got, perhaps, one of the more dramatic answers in U.S. history. What would the response to Gagarin's flight be? According to Sidey the response from Sorensen was that it would be "strong and dramatic." It would be "We're going to the moon."[3]

Six weeks later the answer Sidey had received from Sorensen would be communicated to the American people—and the world—in Kennedy's second address to a joint session of the U.S. Congress in just four months in office. Officially addressing "Urgent National Needs," it was a speech focused on foreign relations, defining friends and enemies, and full of Cold War rhetoric with multiple allusions to the Soviet Union, and declarations of the importance of freedom. It is not until the very end that Kennedy turns to the moon in his oft-cited quote. Again referencing the Soviets and *Sputnik* in 1957, the president declared, "I believe that this nation should commit itself to achieving the goal, before this decade is out, of landing a man on the moon and returning him safely to the earth" (Kennedy 1961). Perhaps not surprisingly, this is very much an answer to Sidey's question about a "national space goal" from six weeks earlier, and of course, Kennedy also follows Sidey's timetable calling for the landing by the end of the decade.

And on July 11, 1969, Americans sat before their televisions and watched the culmination of what became a major part of John F. Kennedy's "New

Frontier" flickering in black-and-white images being beamed to them from the moon as Neil Armstrong took his "one small step." And then, it seems they simply lost interest. Each successive mission had far fewer viewers than the one before it. To many that "small step" was all that was needed to show a victory in the Space Race against the Soviet Union, thus leaving the following Apollo missions only a role of redundancy. As a result NASA could not convince Americans—or the world—to stay all that excited about the moon missions after Apollo 11.

NASA: COLD WAR ORIGINS AND FRONTIER RHETORIC

The origins of NASA are found twelve years before Armstrong's moonwalk and three years before Sidey's questions in the White House. Americans went to bed on October 4, 1957, and woke the following morning to the news that the Soviet Union had placed a basketball-sized artificial satellite into Earth orbit. The flight of *Sputnik 1* (and later *Sputnik 2* with the dog "Laika") would energize what had been a collection of different American space programs, and lead congress to organize NASA. As a result it has always been a relatively simple story for scholars to analyze, and it has always been one driven by Cold War ideology. In the immediate aftermath of the flight of the first *Sputnik*, Congress would organize the Select Committee on Astronautics and Space Exploration in the House of Representatives and the Special Committee on Space Technology in the Senate.[4] The Senate committee would produce the report that would eventually lead to the formation of NASA through the National Aeronautics and Space Act of 1958, which was largely shepherded through Congress by the Senate majority leader and future vice president Lyndon Johnson (National Aeronautics and Space Act of 1958 (as Amended) 1958).

Bringing together both military, academic and civilian groups, NASA was created by combining aspects of the navy's "Project Vanguard," the National Advisory Committee for Aeronautics (NACA), and, shortly afterwards, the Army's Jet Propulsion Laboratory as well as, also from the army, the Development Operation Division of the Ballistic Missile Agency (Grey and Grey 1962; Emme 1966; Shelton 1967).[5] Although motivated by Cold War needs in the first list of "recommendations" from the federal government it can already be seen that the role of NASA is based, at least somewhat, on a frontier mentality. In the 1958 Senate document titled "Recommendations to the NASA Regarding A National Civil Space Program" their mission is defined, in part, as "[a] national civil space research program to explore, study, and *conquer* the newly accessible realm beyond the atmosphere" (National Aeronautics and Space Administration and NASA Office of Program Planning and Evaluation 1959; emphasis added). The intent to "conquer" the "realm beyond the atmosphere" is rhetoric no different than the conquering of the American "frontier." And while NASA was fully aware of the difficulty of such rhetoric, noting that "Space will not submit readily to conquest" as early as 1959, both the Cold War and frontier mythology underpinnings were set for the agency early in its existence (NASA 1959, 3).

Because the original organization of NASA after the Soviet Union launched the first *Sputnik* satellites in 1957, and because in the early 1960s President Kennedy's speeches so frequently turned to the Soviets as a reason for such an endeavor, the consensus among political scientists and historians has been to interpret the Apollo moon missions as part of the Cold War conflict of the period. It is a relatively simple path to follow, after all. The frontier rhetoric, if mentioned at all, is interpreted simply as flowery hyperbole to promote Cold War aims. And those Cold War aims are, at the very least, somewhat obvious. As James Schefter put it, "Kennedy redefined the race" between the Soviets and the Americans in aiming only for the moon and, "winner take all" (1999, 146). The U.S. House of Representatives made their intent specific as early as March 1958 that setting up a new Select Committee, which they claimed was motivated by the

> [c]oncern over the Soviet Union's Launching of the Sputnik space satellite in October 1957 [which] led the House of Representatives to establish the House Select Committee on Astronautics and Space Exploration, the predecessor of the Committee on Science and Astronautics. This select committee was established on March 5, 1958, to study and investigate "all aspects and problems relating to the exploration of outer space."
>
> (U.S. Congress 1958, section 9.1)

As a result of such obviously sign-posted documents as those coming from the U.S. Congress or the speeches of Kennedy, interpretations of NASA, be they in organizational structure and mission selection, have largely been viewed through a prism of Cold War politics. Politics scholar Michael Sheehan sums up such an interpretation succinctly:

> The successful manned lunar landing and safe return of Armstrong and Aldrin in July 1969 represented the achievement of a number of important goals for the United States, both in terms of the Cold War military/technological competition and the prestige and propaganda objectives pursued since the shock of the Sputnik launch. Sputnik had produced profound alarm in the United States and the successful conclusion of the Moon landing set the final seal on the process of recovery and response to the perceived Soviet challenge.
>
> (2007, 53)

Sheehan is undoubtedly correct in proposing that a significant aspect of the organization of NASA, and the decision to go to the moon, were based on immediate political needs directly reducible to the Cold War. In early 1961

Jerome B. Wiesner, who would become Kennedy's science advisor after his inaugural, presented the president-elect with a committee report on the space program, and his conclusions were all Cold War related. Wiesner, a professor from the Massachusetts Institute of Technology, presented any kind of space program as centering on the Soviet Union and a quest for influence among the nonaligned nations of the third world. "First," his report begins, "there is the factor of national prestige. Space exploration and exploits have captured the imagination of the peoples of the world" (Wiesner 1961, 2). Constantly returning to the threats from the Soviet Union, Wiesner concludes, eventually, that this question of international Cold War prestige must be overcome. "Neither NASA as presently operated," Wiesner concludes, "nor the fractionated military space program nor the long-dormant space council have been adequate to meet the challenge that the Soviet thrust into space has posed to our military security and to our position of leadership in the world" (Wiesner 1961, 4).

Kennedy's speeches, and later those of Lyndon Johnson, will reference the Soviets with almost every mention of the moon. As a result the Cold War rhetoric is relatively simple to both establish and to offer judgment upon. California-born historian Gerard De Groot summarizes the entire message of the 1960s American space program in a fashion that captures both this Cold War reasoning and, as a result, the historical moral judgment he brings to it:

> Thanks to Kennedy, Americans were on their way to the Moon . . . The decision was based not on science, but on cold hard politics. [Secretary of Defense] McNamara needed to save the aerospace industry. Johnson wanted to restore American prestige. Congress worried about losing influence in the Third World. Senators wanted fat contracts for their states. Kennedy needed to rescue his image. Everyone wanted to beat the Russians. America would go to the Moon for all the wrong reasons. (2006, 152)

De Groot's judgment, which is sure-footedly written and wholly negative in its conclusions about the lunar missions, is hardly unique. Presidential historian Michael Beschloss concludes in a more measured tone than De Groot does, but it is one that summarizes an entire school of negative conclusions: Kennedy's "decision for an accelerated Moon landing was ultimately a political decision made in terms of cold war strategy. How does it stand up now that the cold war is over? Not well" (1997, 63).

From this interpretation, then, the landing of Apollo 11 did provide Sheehan's "final seal," and there was very little left to accomplish. The Space Race of the 1960s was merely an artifact of the Cold War in this narrative, and as James McDougall would put in his Pulitzer Prize–winning history of the period, the "space program was a paramilitary operation in the Cold War, no matter who ran it" (1985, 174). This seems to fully explain the reasons why the missions that followed Apollo 11 had significantly reduced popularity quite effectively. To put it more succinctly, the main interpretation is that the motivation for the 1960s Space Race was the geopolitics of the Cold War and was a dubious justification at best. But, if nothing else, scholars were able to then conclude that the Americans had at least won it.

The preeminent historian of the lunar missions, John Logsdon, at least hinted at why Apollo is included in this book, however. "Kennedy's most important reasons for wanting to formulate a response to Gagarin were linked to foreign policy," he writes, "especially to maintaining the American position as the leading world power" (Logsdon 1970, 108–9). This chapter does not argue the point that Cold War considerations were significant and central to NASA's mission in the 1960s. It does, however, argue that other considerations about the American mythscape—as evidenced in the rhetoric used to support the mission—must be considered too.

The number of works espousing the notion that the moon landing was merely an extension of Cold War geopolitics with the addition of some frontier rhetoric is exemplary of this geopolitical interpretation. The resulting placement of any serious examination of frontier mythology is thus firmly, and ironically, on the periphery. The majority of histories written on the period agree with Roger Launius's conclusion that the "primary goal" of the Apollo missions and the 1969 moon landing was simply to "establish [American] technological preeminence over rival nations and little else" (2004, 23). Yet this Cold War thesis tends to ignore that frontier rhetoric, or merely consider it part of a selling point with which the Cold War was fought in space. However, with post–Cold War declassification of documents the geopolitical reasoning for the Space Race seems less than secure. As Hirotaka Watanabe has noted after reviewing recently declassified documents, at the very least one must consider that a "relaxation of the tensions in the Cold War was the purpose of Project Apollo," and any other reading, regardless of how tidy its conclusions might appear, is likely profoundly overly simplistic (2009, 44–45). In short, simply examining a Cold War Space Race is not the only way to look at the period, as was noted even at the time. Indeed, frontier rhetoric had been part of the NASA institutional culture and its public face since the beginning. In Congressional Hearings in 1960 they had accepted that "we yield to the urge to explore that is an American heritage" (NASA 1960, 159). By 1965 NASA had fully accepted the mythic frontier underpinnings of their overall project, proclaiming that its missions were "exploration in the truest and most romantic sense" and that space was, therefore, "the most recent of these 'last' frontiers" of such exploration (NASA 1965, 2).

While never a mass movement, some segments of the American population had been supporting the idea of a large-scale space program for years based, in part, on NASA's willingness to utilize frontier mythology in promoting itself. At the same time, as American professor of Public Affairs and NASA expert Howard E. McCurdy has put it, in the 1960s those "space advocates found a ready supplement in public anxiety about the Cold War" (1997, 53). But Cold War rhetoric, proximate international relations motivations, and Machiavellian politics do not explain the entirety of NASA in the 1960s. Even in the 1960s this was noticed, although historians since that time have largely ignored other explanations because their own conclusions are, at least in part, reasonable. Also, such explanations allow for primarily discordant interpretations in a period full of conflict and open to such interpretations. Although NASA was given to statements such as "[T]he new frontier has been opened. A new domain—outer space—awaits conquest by man," Cold War political science and history scholars have found little of such rhetoric to consider positively (NASA 1960, 159).

Yet just a day after Kennedy's declaration of "before this decade is out," the *New York Times* already knew full well that the Cold War implications of the Space Race was part of the purpose but editorialized beyond that saying that such a race with the Soviets was "only secondary to the main purpose," which was "unquenchable thirst for exploration of the unknown" (New York Times Editors 1961, 32). This kind of interpretation, which Howard McCurdy described as an attempt to "reactivate the vast migrations that molded America in the past" is both naively optimistic and based on a frontier mythology that was never actually part of the real historical past (1997, 140). For McCurdy, this is the main problem of such myths, and for this book, it remains their main strength.

Yet, like that version of the myth from the previous century, NASA's mission in the 1960s seemed, for a time, equally inevitable. Writing just after Kennedy's "Urgent National Needs" speech the American science writer Jay Holmes did not even try to present an image of journalistic objectivity when he noted that for "thousands and perhaps millions of years to come, school children will have to memorize a date in the latter 1960s—probably during the second administration of John F. Kennedy—the day men first set foot on the moon" (1962, 17). In a book about the just emerging lunar program subtitled "The Enterprise of the 60s" Holmes was aware of the Cold War aspects of the Space Race, and mentions the Soviets with great frequency. At the same time, however, the moon program and eventual lunar landing would far outstrip such mundane realities as the contemporary international politics of the time and be "like Columbus' arrival in San Salvador." It would thus "be remembered in history as a symbol of civilization bursting through a barrier" (Holmes 1962, 17). While obviously part of contemporary geopolitics, Holmes still presented an aura of inevitability and historical necessity to the lunar missions. The geopolitical reality remained the fear that "human civilization destroys itself in a nuclear war" before the entirety of the universe could become a new frontier. If such a catastrophe was avoided, however, then "when it becomes technically feasible, men will explore the planets, the stars and even other galaxies" (Holmes 1962, 251). Henry S. F. Cooper, Jr., writing months before the landing of Apollo 11, put it more bluntly: "man's destiny must somehow be worked out among the stars," he wrote and then said of the personnel at NASA, "[this] notion may be part of what keeps them all going" (1969, 19).

For a certain part of the American public, the sense of inevitability of an eventual lunar mission had been building for some time, and significantly predates the arrival of Kennedy to the White House. A staple of science fiction since the nineteenth century imagined lunar missions had long been part of American popular culture. More significantly, however, would be the arrival of such ideas in the larger-circulation press. Beginning in 1954, and running through eight issues, one of the largest magazines in the United States, *Collier's*, had offered coherent plans for eventual exploration of space, the moon, and Mars beginning in the spring of 1952 and running through spring 1954 (Collier's Editors et al. 1952). With a cover story for their March 22, 1952, story blazing the headline "Man Will Conquer Space *Soon!*" essays were offered during the run of the series from German expatriate and science writer Willy Ley, American astronomer Fred Whipple and, most significantly, former Nazi rocket designer Werner von Braun. The *Collier's* articles came to dominate visions of space exploration for not just the period, but for decades to come. Indeed, as T.A. Heppenheimer has described it, the *Collier's* articles offered a plan for "cargo rockets such as the Saturn V and the space shuttle, astronaut moon landings, a space station, the eventual flight of people to Mars" (1999, 6). In short, the *Collier's* series offered all the "ideas that would dominate NASA's projects and plans" for the next fifty years (Heppenheimer 1997, 89–90). And those plans were organized around von Braun's ideas and, significantly, emphasized the Cold War aspects of space exploration he had used to sell the series ideas first to *Collier's* and, later, to NASA itself.[6]

Thus, the Cold War paradigm, while both reasonable and insightful, remains only one way of examining the lunar missions, the success of NASA in the 1960s and early 1970s, and the precipitous decline in the popularity of the space program after that period. But Jay Holmes's work approached a different perspective that was inherent at the time, but missing in political and historical analyses since the Apollo missions, was brought to an early end. While acknowledging the importance of the Cold War, there remains a strand of thought that suggests NASA's work on Apollo was of a significantly larger scale than simple geopolitics. Since the end of the Apollo lunar missions this perspective has largely been deemed to be part of the propaganda support structure build to support significantly increased budgetary outlays. But it is possible to take this perspective seriously in its own right, and examine it as something more than simply propaganda and rhetoric, but as mythic resonance. When, in 1960, NASA administrator George Low told a group of newspaper editors that "man is destined to play a vital role in the exploration of the moon and of the planets," he was providing more than rhetoric for a wholly Cold War–motivated activity (1960, 2; this quote and the following two citations are discussed in Byrnes [1994, 60–64]). NASA was promoting, regularly and intensely, the idea that the exploration of space "was part of America's future destiny," and that this destiny was "inevitable" (1960, 160–161). When offering such images of space exploration as "the certain destiny of man," as one 1969 NASA document did, then confining an analysis of NASA in the period to little more than an offshoot of Cold War ideology verges on oversimplification (NASA and Anderton 1969, introduction).

Moving beyond Cold War explanations is a decidedly more culturally nuanced and problematic process, one that most certainly does *not* end in American success by landing on the moon. Rather, it is a complex set of mythographic links presented by President Kennedy and others that, when Apollo 11 landed on the moon, doomed the project in a host of unrealistic mythological expectations. Specifically, if the American mythos grants to the notion of the frontier, as Henry Nash Smith put it, "grandly and abstractly, a place where afflicted humanity raises her drooping head," the NASA astronauts found no such location in their travels (1971, 129). Rather what they found could best be described with the two words spoken by the second person to walk on the moon, Buzz Aldrin. Climbing down the lunar module ladder and looking at the moonscape before him, he said, first, "[B]eautiful view!" It was a statement that could have been produced by English colonists of the seventeenth and eighteenth centuries looking at Massachusetts Bay or the Chesapeake in Virginia. Farmers could have uttered it in the early nineteenth century as they homesteaded in the Old Northwest or ranchers in the middle nineteenth century in Montana. It was, in short, the appropriate statement to make by a man selected to appear as a contemporary version of the pioneering spirit that abounds in American myth.

But then he declared, honestly, "[M]agnificent desolation" (NASA Historical Reference Collection 1969, 382).

TURNING FRONTIER TERRAIN INTO FRONTIER TROPE

From its origins NASA officials offered frequent statements relating their mission to the general myth of the frontier they believed to be understood by all Americans. Prior to President John F. Kennedy declaring the lunar program in 1961, however, only a third of Americans polled favored such a project. By 1963 that percentage was more than two-thirds in support (Gallup 1720, 1972). That change cannot be wholly ascribed to prevailing Cold War ideology, nor can the many previous instances of NASA officials publically comparing space to "the frontier" offer a conclusive reason. The change was in the method in which a rhetorically gifted president offered "the frontier," manipulated how it was understood, and then applied it to his own space program. In reusing, reworking, and redefining "the frontier" through the early 1960s, Kennedy offered the term as a mythological trope that would greatly benefit his Presidency, but ultimately doom any possibility of long-term success for the lunar missions he employed the frontier to promote. Accepting the Democratic Party's nomination for the presidency, Kennedy, standing in Los Angeles, specifically evoked the idea of the frontier not as a place, but as a useful metaphor. This was the first shift in the trope of the frontier that Kennedy would eventually use to support NASA's Apollo program. He said,

> For I stand tonight facing west on what was once the last frontier. From the lands that stretch three thousand miles behind me, the pioneers of old gave up their safety, their comfort and sometimes their lives to build a New World here in the West. . . But I tell you the New Frontier is here, whether we seek it or not. Beyond that frontier are the uncharted areas of science and space.
>
> (Kennedy 1960)

Kennedy specifically used the "frontier" as a metaphor. Eventually this metaphor was extended as an all-encompassing trope when Kennedy begins to speak of the United States' lunar goals. It is common for historians to paint Kennedy's use of the Space Race and NASA in Cold War terms, but this is an oversimplification. Kennedy will repeatedly evoke the frontier during his presidency (indeed, his term will be defined by it), but almost always in the sense of metaphor or trope. While historians of NASA in the 1960s have tended to emphasize what Walter McDougall calls an attempt at "win a battle of image making in a total Cold War," the result of that thesis has been to overlook the frontier rhetoric of the American civilian space program in the same period (1985, 304). This is a perfectly valid interpretation because, to those studying NASA, the immediate drop-off in both the popularity and funding can be easily explained by the simple fact that in landing on the moon in 1969 the United States had, for lack of a better term, "won" the Space Race.

However, an examination of the frontier rhetoric and frontier myth-making used by Kennedy and others offers a more complex reasoning and must be taken seriously on its own merits and by its own logic (Fulmer 1986). In using the long-standing frontier ideology of American culture in the promotion of an all-out effort to "land a man on the moon and return him safely to the earth," the project was left with a mythological need required by such rhetoric that could not be fulfilled. When Buzz Aldrin called the moon "magnificent desolation" he also, unwittingly, noting that this new frontier—unlike the myths that it was based on—would not and could not be the virgin land of which the myths spoke. American Studies scholar Charles Sanford wrote of the disillusionment over paradisiacal frontier myths in the twentieth century. Noting that "the most hallowed soil in America from its long-time connection with the Edenic dream" still existed well into the middle of the century, he also argued that such dreams were now fully tempered with the idea that such beliefs were no longer possible as a reality. If the Apollo astronauts stood for anything, in Sanford's version of such myths, it was that mythical and metaphorical place

> where the self-reliant pioneer farmer, driven ever westward in quest of the New Jerusalem, had lost his momentum, become weighted down with mortgages and surrounded by skyscrapers—the Mississippi Valley, "the valley of democracy," became . . . a Wasteland.
>
> (1961, 255–256; original emphasis)

Thus, when Aldrin proclaimed the desolation of the lunar surface he only confirmed the impossibility of such mythic dreams to the American public. That the mythic frontier employed to support the NASA lunar mission demanded and fully required that what the astronauts discover would be somehow akin to the mythic frontier of the past—a virgin land or the possibility of a New Jerusalem—can be easily overlooked by scholars willing to concentrate on the parallel Cold War rhetoric of the period. But an examination of that frontier rhetoric, and what it demanded, leads to another conclusion about the loss of momentum for the space program. Rather than Apollo 11 signifying a Cold War victory in reality is singled a mythological failure. The rhetoric employed to gain support for the lunar missions was a rhetoric that demanded a certain quality in the terrain it explored and the moon, for all its harsh beauty, could never be the terrain demanded of it by myth. The result is that the Apollo missions had gained momentum by employing a mythological structure that, in the end, was so loaded against success that the loss of NASA's popularity was mythically unsurprising.

Thus, an examination of such rhetoric—and the myths it supported and that it was supported by—must begin with the greatest purveyor of such images: John F. Kennedy.

SPEAKING MYTHICALLY: KENNEDY AT RICE UNIVERSITY

There are few presidential speeches that concern themselves wholly with space. Indeed, other than those few resulting from tragic accidents such as the *Challenger* and *Columbia* disasters, it is John F. Kennedy's (1962) speech at Rice University in Houston, Texas on "The Nation's Space Efforts" that stands out as the only major address by a sitting president on the space program (Kennedy 1962). Kennedy's (1961) "Urgent National Needs" speech is a mainstay in television documentaries and news shows because of its short, declarative "before this decade is out" statement but the Rice University speech is a far more complete rendering of Kennedy's position on the space program and lunar mission. And, while scholars often quote from it, they do so in usually selective fashion to highlight the Cold War aspects of the speech. Space historian Andrew Chaikin's sprawling work *A Man on the Moon*, purports to offer a nearly complete history of the

Apollo Program, and does begin with the Rice University speech. (Chaikin 2007).[7] Yet Chaikin uses that address merely for entrance into his history through a tool of nostalgia, and offers comparatively little on the speech beyond that introduction. If one form of analysis sees the NASA missions of the 1960s operating between an axis of domestic politics and the locations of NASA centers and moving from there to the international politics of the Cold War, Chaikin offers another viewpoint. What he offers for the same period, through an extensive series of detailed interviews with surviving astronauts, ground controllers, and other government employees, is a story not of political success, but along an axis of either personal or organizational success. Like the Cold War narrative, however, Chaikin ends with exactly that: success. When Armstrong and Aldrin had landed on the lunar surface the mission has been accomplished—be it in political or personal terms—and the cost of future missions is simply no longer supportable. Even with his exhaustive interviewing and deft narrative abilities, however, Chaikin does not dally overly long on the broader cultural meanings of the lunar missions. And after that briefest of mentions to Kennedy's Rice University address, Chaikin turns to the established Cold War narrative as well, declaring that "[s]pace was the new battleground of the Cold War, and the Soviet Union was winning" (2007, 1).

With either the Cold War narrative, or the narrative of personal success typified by Chaikin, the story ends with the success of the lunar landing. With each narrative the reasons for ending the Apollo missions prematurely differ, but the logic of the ending is not questionable. There are to be no moon bases, no Mars missions, and, after 1972, no human beings outside of low Earth orbit. But, regardless of which narrative that is adopted, there is at the very least the unrivalled success of Apollo 11.

Yet, Kennedy's address at Rice University, aside from being only occasionally about either the Cold War or a perceived Space Race, stands out as a speech designed specifically to resonate mythically. American rhetoric scholar John W. Jordan has claimed that as a result of Kennedy's address later presidential speeches "evinced a continued desire to incorporate romantic and transcendent elements into space exploration imagery" but that most failed where Kennedy had succeeded (Jordan 2003, 225). Kennedy's address offered both a more complete statement about space exploration and, at the same time, incorporated a significant retelling of frontier mythology, what counted for popular frontier history, and simultaneously used a mythic American past as a way to present both a present and a future for his audience that made the lunar mission appear both necessary and inevitable. As a result, for Jordan, the lack of mythic integrity of post-Kennedy space policy speeches is in a large part due to the very effectiveness of the Rice University speech.

A detailed analysis of Kennedy's speech can show exactly how such "romantic and transcendent elements" could further Kennedy's space goals and do so by using myth is a politically purposeful fashion. It is also the single most important speech on space policy ever delivered, and that is because of its effective mythical underpinnings. Dwight Anderson has argued that often Kennedy presented "clever misreading of American texts" resulting in a "presidency of words in search of events commensurate with the gravity of his language" (1988, 209). With the Rice University speech, however, Kennedy uses myth and rhetoric to create events toward which his language can then aspire. If myths, themselves, work as William G. Doty has described, as a "network" of "culturally important stories conveying, by means of metaphor and symbol, graphic imagery, and emotional conviction . . . foundational accounts of the real, experienced world," then Kennedy's speech on September 12, 1962, was as much a mythmaking exercise as it was an exercise in political rhetoric (1980, 531). At the time Kennedy's call for a lunar mission was far from certain as a political and popular goal. Reporter Hugh Sidey's *Life* magazine had noted in a cautiously positive editorial that not only were most Americans against the lunar program after Kennedy's "Urgent National Needs" speech but primarily so because of the price associated with the effort (Life Magazine Editors 1961, 54). What he had to accomplish at Rice University, as James L. Kauffman described it, was offer a "political, scientific, military, and economic justifications for sending a man to the moon" (Kauffman 1994, 5). To do this was to do more than merely provide a series of pragmatic justifications; Kennedy had to "depict Project Apollo as a structurally coherent narrative" by uniting the entirety of the NASA lunar program within a confines of a single story (Kauffman 1994, 5). The story Kennedy's Rice University speech provided was one supported far more coherently by American frontier myths that it was by Cold War rhetoric. Kennedy offered, then, what Mircea Eliade describes as a narrative of history for the New World, generally, but specifically for an American context. It was a speech that presented a history that turned "back and [found] their *primordial history*, their 'absolute beginnings,'" while at the same time presented a "future not only better, but beatific" because of the space program (Eliade 1969, 89; 98; emphasis in the original). That past, the "primordial history" of the United States, as well as the "beatific" future was combined through the myth of the frontier. As Kauffman describes it, in Kennedy's speech "the frontier narrative stood as the most powerful justification for a manned moon mission" (1994, 29). This frontier mythology, which was only part of the few minutes devoted to the "Urgent National Needs" speech the year before, is the central motif of Kennedy's address at Rice University. And, if Kauffman's assertion that this speech represents a significant moment of the "crucial period" of the American Space Program, it is also true that the program as a whole is one that is "steeped in myth [and] to understand the myths is to gain a better understanding of the origins and evolution of American space policy" (Kauffman 1994, 1).

Standing in the center of the university's sport stadium and speaking to forty thousand people on the occasion of a donation and sale of land by the university to NASA for what would eventually become the Johnson Spaceflight Center, Kennedy's rhetoric brought together a powerful network of metaphors, symbols, images, and emotions specifically designed to win support for the lunar missions among the American people. As Janice Hocker Rushing has noted, to do this required that he relocate the American frontier narrative into outer space, which in and of itself required a fundamental change in the implications of what the term *frontier* meant. Such transitions are not new, according to Rushing. It is merely that the transition of the frontier from "unknown regions from sea to land" was moved by Kennedy so that "we seem now to be excited more by the stars than the range" (Rushing 1986, 266). Because outer space is infinite and empty, Rushing argues, it was necessary to offer a new kind of map of the frontier. Additionally, because this new frontier could only be, in the traditional mythic language of the narrative, "conquered" through the application of technology largely applied from ground stations making the astronauts mere "passengers" the notions of frontier heroism had to be altered as well. In short, Kennedy had to tie his lunar ambitions, which largely did not precisely fit into a "frontier" narrative to a new version of that narrative created specifically to "sell" the project.

Exactly how his speech—likely written primarily by Kennedy advisor Ted Sorensen—accomplished this is a study both of the masterful rhetoric the president was capable of and of the mythic symbols of the frontier that American culture found so appealing. While supporters of manned space flight had for years promoted the objective, at the Rice University speech Kennedy had, in just thirty or so minutes, offered what biographer Ralph G. Martin described as "the real birth of the American Space program" (1983, 340). The speech itself, then, can be seen as the first extended statement of both the complexities and the importance of the efforts of the American government to develop a space program. The rhetoric scholar John W. Jordan has carefully examined the speech in ways that mirror the use here. For Jordan, Kennedy's speech "rested on overlapping rhetorical constructions of space, time, and the public as pioneers" and thus invited his audience—and by extension the American public—into a position of support for the lunar goals (2003, 214). The overall result of the speech, which I describe as mythic, Jordan sees as a rhetorical tour de force in its own right. He describes the organization of Kennedy's address, and its underlying rhetorical strategy, this way:

> The Rice University address is characterized by a transcendent rhetoric that utilizes three major strategies: a characterization of space as a beckoning frontier; an articulation of time that locates the endeavor within a historical moment of urgency and plausibility; and a final, cumulative strategy that invites audience members to live up to their pioneering heritage by going to the moon.
>
> (Jordan 2003, 210)

For Jordan the end result is an effective statement of space policy that has significantly affected later statements by future presidents by existing as a comparative text. Jordan's interest, as a rhetoric scholar, is not surprisingly within the field of rhetorical textual analysis. Thus, while noting that other scholars have attempted to trace a decline in NASA's popularity after the Apollo lunar missions to Cold War politics or other causes, Jordan claims that "Kennedy's rhetoric . . . later became a problem for those [presidents] who sought to emulate it" (Jordan 2003, 226). In part this is simply because Kennedy's time in office has been, among recent American presidents, the most mythically interpreted by the American electorate (Felkins and Goldman 1993 offer a quantitative analysis of Kennedy's political rhetoric and the reasons for its acceptance by the American public). At the same time, Kennedy's rhetoric is at once soaring and carefully attached to the mundane, looking to the past and imagining a future, making the Rice University address almost impossible for later politicians to imitate. And, Jordan notes, it can

> therefore be understood as an attempt by the president to define an ambiguous situation (America's space program) in terms of shared communal values (pioneering exploration) with the express purpose of urging that community immediately to take up the challenge (going to the moon).
>
> (2003, 213)

In effect what Kennedy offers in his address is a two-pronged mythic restatement of frontier history and frontier geography. He first presents a new version of history that, while superficially familiar, affects a connection between the present and the past specifically designed to articulate a sense of historical urgency in his listeners. The history he presents becomes Eliade's "primordial history" which asks of the audience to find "the desire to start out again," as is always the case in such narratives, Eliade argues (1969, 89). And that new beginning, that desire to fuel the future rhetoric, happens to the audience through Kennedy's speech by supporting the Apollo mission. Kennedy then offers a starting point for that new beginning by offering a version of the geography of outer space firmly connected to terra firma and designed to hide the "infinite" scope of outer space in favor of a smaller map of just the Earth and the inevitable and necessary frontier of the Moon. His historical and geographical reimagining leaves the lunar mission placed as both an obvious and a necessary step.

Kennedy begins his address with a series of statements that, when combined, begin to point to his sizable lunar goals. "We meet at a college noted for knowledge, in a city noted for progress, in a State noted for strength," he declares at the very beginning of his remarks (Kennedy 1962, 4).[8] The organization of his nouns moves from the smaller to the large: "college" to "city" to "state." This progression intentionally draws his listeners to move toward larger categories, but first Kennedy presents a series of contrasts: "And we stand in need of all three, for we meet in an hour of change and challenge, in a decade of hope and fear, in an age of both knowledge and ignorance" (1962, 4). These contrasts serve to allow Kennedy to acknowledge the very breadth and contentious nature of actual lunar missions. The scope of "space" as a concept, of "eternity," and the "infinite" are all issues that the president must deal with rhetorically in his address. He does this first by offering a new version of history designed to make the events of the past far more immediate to his listeners.

"No man can fully grasp how far and how fast we have come," he notes in acknowledging the immense size of his themes. But then he offers his audience a way to cope with such large-scale ideas by offering a firmly remade and grounded version of world history: "but condense, if you will, the 50,000 years of man's recorded history in a time span of but half a century" (Kennedy 1962, 6). By reducing the scale of all human history Kennedy has also reduced the scale the present moment. This will become significant when, as a result, he can make his support for Apollo seem at once obvious and inevitable. For the moment, however, it allows him to offer a past to his audience that appears much closer to them than would otherwise be the case. With fifty thousand years of history in merely fifty years, Kennedy notes,

> [A]bout 10 years ago, under this standard, man emerges from his caves to construct other kinds of shelter. Only five years ago man learned to write and use a cart with wheels. Christianity began less than two years ago. The printing press came this year, and then less than two months ago, during this whole 50-year span of human history, the steam engine provided a new source of power.
>
> (1962, 6)

This strategy allows Kennedy to put into a very near perspective such things that appear to the average American to be part of a distant past. And even the more recent past becomes much closer: "Only last week did we develop penicillin and television and nuclear power," he continues his thrust of rhetoric bringing the past into a position where its relevance to the present seems more obvious. This is a mythic restatement of history that Kennedy uses to overcome the traditional American tendency to ignore history. William G. Doty, in noting the tendency in American myth to present a "dichotomy between the past and the present, with the assumption that only the present really matters," marks the difficulty in historical presentations that resonate with contemporary audiences in the United States (1995, 193). Kennedy sidesteps this problem by collapsing all of history to a period so short as to seem recent to his listeners. Then, as Doty notes about myth functioning to "geographize the natural contours of creation . . . [and] link meanings . . . synthesis, linking, holding together," Kennedy adds to his history both the present, and a new geography of space as well (1995, 194). In doing this, he also implies an equally immediate future by noting that the NASA probe Mariner 2 was at that time approaching Venus: "and now if America's new spacecraft succeeds in reaching Venus, we will have literally reached the stars before midnight tonight" (Kennedy 1962, 7).

Collapsing both outer space geography and history into an arbitrarily short area and short period works for Kennedy because it makes the past seem part of the present, and makes the infinity of space seem much closer. And it makes his audience feel connected to both space and the past more personally. And, by foreshortening the period, Kennedy also implies a strict and inevitable narrative logic to his retelling of the past, while suggesting an inevitability of the exploration of space in the present and future. He calls it a "breathtaking pace" and offers, "surely the opening vistas of space promise high costs and hardships, as well as high reward" (Kennedy 1962, 8). But by bringing both the recent and the distant past into such sharp relationship with the present Kennedy's fifty-year analogy of human history makes the past seem inevitable and, thus, makes the future he is presenting seem equally so. Then by immediately attaching to his historical narrative his mention of Mariner 2 coasting past the planet Venus Kennedy has attached to his seemingly inevitable historical narrative a new narrative of the present and the future, with a new map of precisely what this present and future relates to.

At the same time the address reimagines a past that has the option of placing the present mythically outside of history. The empirical historical past becomes a nonhistorical mythical past in Kennedy's rhetoric, placing the present in a mythical position, and therefore in need of a mythical response. By presenting the past in a mythical timeframe, Kennedy offers his listeners the opportunity, as Mircea Eliade put it in a different context, the knowledge that "the individual's projection into mythical time [does] not occur, of course, except at essential periods," thus suggesting that the present Kennedy is speaking about is such an "essential period" and that his listeners become people taking part in "important acts," which are "projected into mythical time, into the moment when the paradigmatic voyage took place" (1954, 35–36).[9]

Kennedy then begins to place a historical obligation on his audience, reminding them that the past he has presented is one that had provided a land such as "this city of Houston, this State of Texas, this country of the United States," all of which had been "conquered by those who moved forward—and so will space" (Kennedy 1962, 9). The history Kennedy provided thus became linked to the frontier tropes and myths that his audience fully understood. "Space," like the trans-Mississippi West before it, was a place to be conquered. And such a place could be conquered as Americans had done such things before. He reminded his listeners of exactly what they believed such a history to mean, telling them that "those who came before us made certain that this country rode the first waves of the industrial revolutions, the first waves of modern invention, and the first wave of nuclear power" (Kennedy 1962, 12). The wave-driven nautical imagery was intentional as space was to be, just a few moments later in the address, "this new ocean," but for now he assured his audience that with such a history behind them "this generation does not intend to founder in the backwash of the coming age of space" (Kennedy 1962, 12). Indeed, if Henry Nash Smith was even slightly correct in proposing that a notion of American Empire could "embody two different if often mingled conceptions" of an "empire as command of the sea" as well as one of a "future society occupying the interior of the American continent," then Kennedy's nautical rhetoric nicely forces the intermingling more completely (1971, 12). Speaking of the city of Houston, again referred to in metaphorical scale as "this city and this State, and this region," Kennedy could note that "what was once the furthest outpost on the old frontier of the West" of that inland empire "will be the furthest outpost on the new frontier of science and space" (1962, 25). The former inland empire was set, through Kennedy's rhetoric, to become the space-going port of "this new ocean," bringing together both kinds of empires into one, single project: Apollo.

Significantly, this is a historical retelling that allows the technological advances of the past to become part of an ethical narrative of progress. This "new ocean" is one that, as Charles Sanford traced the concept to the middle of the nineteenth century, "identified technological progress with moral and spiritual improvement through the agency of the factory and a benevolent paternalism" (1961, 171). The purpose of the Apollo program is, for Kennedy's rhetoric, moral progress as much as spatial advancement. For Kennedy Americans will "set sail on this new sea because there is new knowledge to be gained," but they will do so through the moral certitude of the centrally organized and technocratic Apollo program. But the address only works because, in the same fashion that he reduced fifty thousand years of human history to a mere fifty, Kennedy continues to relate the lunar mission to the everyday and mundane as well as previous historical achievement. "But why, some say, the moon?" he asks, following it with a series of historical firsts. "Why climb the highest mountain? Why, 35 years ago, fly the Atlantic?" Kennedy need not have reminded his listeners that it was just a decade earlier that the summit of Mount Everest had first been reached or that the first solo flight to have crossed the Atlantic in 1927. But then he added, "[W]hy does Rice play Texas?" referring to the American collegiate football game and source of a serious statewide rivalry (Kennedy 1962, 15). In part a lighthearted quip, but at the same time, when placed alongside Edmund Hillary and Tenzing Norgay's feat on Everest or Lindbergh's landing in France, Kennedy offered, in effect, the idea that his plans for Apollo were at once historically significant and simultaneously mundane. That what seemed almost impossibly difficult was, for a nation built on the history of Western pioneering expansion, merely the next obvious and necessary step.

If Frederick Jackson Turner had presented a nineteenth-century social-scientific version of inevitable Manifest Destiny across the frontier continent and then worried that "never again will such gifts as free land offer themselves" Kennedy was representing both the image of that inevitability and a new version of "free land" in space (1999, 40). For his audience the "gift of free land" was beckoning above their heads and took only American know-how to conquer. Thus, when Kennedy presents his most quoted statement from the address he is presenting both a version of historical inevitability and, at the same time, mythic necessity:

> We choose to go to the moon. We choose to go to the moon in this decade and do the other things, not because they are easy, but because they are hard, because that goal will serve to organize and measure the best of our energies and skills, because that challenge is one that we are willing to accept, one we are unwilling to postpone, and one which we intend to win, and the others, too.
>
> (Kennedy 1962, 16)

When quoting this at all scholars tend to emphasize the "we intend to win" phrase. But, like his address on "urgent national needs" and his acceptance of the Democratic nomination for the presidency, Kennedy is really speaking mythically. The Space Race and the Soviet Union are important which is why so many historians have concentrated on it. But also important—centrally so as much as the Cold War—is the rest of the mythic structure. At Rice Kennedy was, as John Jordan put it, offering to "construct for his audience a new motivation for the lunar mission that promised to raise them above conditions of uncertainty and place them at the beginning of what he promised would be the next and greatest chapter in the history of human exploration" (2003, 214). Kennedy was presenting a future based on something other than Cold War geopolitics: instead, it was a future based on hope. This motivation for moving forward evoked the American past mythically and specifically. It is impossible, then, to examine either Kennedy's speech or the lunar mission it supported without realizing, as the American aerospace historian T. A. Heppenheimer has put it, that "in understanding Apollo, one must take note of this element of hope" (1997, 185).

Nowhere is this more specifically made true than when, early in the address, Kennedy recounts William Bradford, governor of Plymouth Colony in seventeenth century colonial Massachusetts. The rest of the Rice speech—and, indeed, the majority of presidential references to the "frontier" of space in general—works within the framework of nineteenth century frontier tropes. But, when Kennedy, recounting the message of Pilgrim New England, notes that "all great and honorable actions are accompanied with great difficulties," he is figuratively setting the rhetorical frame for the rest of the speech, and it is a far earlier version of that mythical "frontier." The lunar missions will have those "great difficulties," but at the same time, like the Pilgrims of mythic history, they will be "overcome with answerable courage" (Kennedy 1962, 10). And when Bradford recounted the difficulties of the new colony in Massachusetts in *Of Plymouth Plantation*, his tone was one that could never be met by the astronauts of Apollo. Bradford could claim, as Kennedy recounted, that "all great and honourable actions are accompanied with great difficulties, and must be both enterprised and overcome with answerable courages" but, just a few short years later Bradford could note that "the people of the plantation begane to grow in their owtward estats, by rea[son] of the flowing of many people into the cuntrie" while NASA could only land but two men per trip on the surface of the moon (1908, 31; 503). There would be no growing estates, full of "diverse cornfeilds, and litle runing brooks, a place (as they supposed) fitt for situation" (Bradford 1908, 128).

At the same time in Kennedy's brief mention of Bradford another shortening of history is accomplished. With Bradford, who recounted the first Thanksgiving banquet that every American schoolchild learns about, with "water foule" and "wild Turkies, of which they tooke many, besids venison, etc." And "Indean corne" leading the colonists to "write so largly of their plenty hear to their friends in England, which were not fained, but true reports," Kennedy has offered a mythically resonant version of history that cannot, simply or easily, be redeemed by a successful landing on the moon's surface (Bradford 1908, 162). More than an historical reference, Kennedy is providing a metonym for the entirety of his mythic message. Bradford, in his diary is but one early English settler of North America that provided in their surviving writings the basic structure that would become the American myth on which Kennedy is attaching his lunar goals. The rest of the speech, almost exclusively about those lunar goals, will continually reverberate with this early reference to Bradford and, through him to many of the founding documents of American mythology.

And here Kennedy is not merely using existing myths known by his audience but is actively engaged in making those myths into something different. Bringing Bradford forward as a metonym for the rest of the national mythos he is using, Kennedy is not merely presenting a mythical past, but offering a new, mythical future as well. William G. Doty notes that while "myths establish conservative benchmarks," in many cases, he also adds precisely the mythical notion that Kennedy is here undertaking, offering that "but at the same time they anticipate *forms of the future* as they determine and shape ideals and goals for both individual and society"(2000, 26; original emphasis).

All that Kennedy needs for his mythical speech to work over the long term, then, is that the environment of space should match, at least generally, the environment described by those seventeenth-century colonists and explorers. And that, as mythical heroes as well as historical personages, his astronauts must be the modern-day equivalents of Bradford in Plymouth, or John Smith in Virginia, or any of the others who provided the images of the virgin, frontier continent upon which the frontier mythology was built. While the seventeenth- and eighteenth-century frontier was as much about "conquering" as would be the mythic nineteenth-century version, its emphasis was on conquering an alien but potentially Edenic environment. It would be this mythic conception of the environment of the New World that, as compared to the environment of outer space, would prove to be the most difficult mission for NASA in the 1960s.

BRADFORD AND THE NEW WORLD AS METONYMIC SIGNIFIER FOR APOLLO

When discussing their motivations for joining NASA 1960s astronauts tended to evoke nineteenth-century mythic interpretations of pioneering impulses and the Western movement in U.S. history. The original seven Mercury astronauts were especially adept at such mythologizing. In a book they published together called *We Seven* they described their work in terms consistent with myths of nineteenth-century Western expansion. Astronaut Scott Carpenter put it bluntly, saying, "[S]pace is a fabulous frontier." Gus Grissom offered that he felt a "spirit of pioneering and adventure" and that "had I been alive 150 years ago" he would have helped to "open up the West," while Alan Shepard, the first Mercury astronaut to be launched into space, explained his decision to join the space program as the "urge to pioneer." John Glenn suggested that the astronauts were attempting to "blaze a trail," and as part of NASA were up for the job to "be the pioneers of this new age" (Carpenter et al. 1990, 346, 64, 58, 28, 327; see also Kauffman 1994, 32–44). In an article written for the ever-present *Life* magazine Glenn went even further, claiming the role of NASA was to take part in the "great exploration of all time" (1961, 46).

This nineteenth-century version of western expansion into space became very much a dominant rhetorical theme after Kennedy's speeches. The Space Race was going to be really a nineteenth-century "new frontier" in as many ways as possible, and if the technological realities got in the way they would easily be bypassed. Kennedy had provided a new history of the frontier leading, inevitably, to the new frontier of space. In doing so he offered a new geography of such exploration as well, making the moon the logical next step for the frontier ideology of America. The former German rocket scientist and then head of the Marshall Spaceflight Center in Alabama Werner von Braun explained the explicitly nineteenth-century and logos-centric logic succinctly to Congress less than a month after Kennedy's Rice University address:

> Everyone knows what the moon is; everyone knows what this decade is; and everyone can understand an astronaut who returned safely to tell the story. An objective so clearly and simply defined enables us to translate the vague notion of conquering outer space into a hard-hitting

> industrial program that can be orderly planned, scheduled, and priced out. It establishes a sorely needed, firm, nonvacillating goal
> ("Von Braun Testimony on Space Plans," 1963).

The "orderly planned and scheduled" frontier ideology would be repeatedly highlighted by NASA scientists, the White House press office, and the astronauts themselves. NASA's general council, John Johnson, would call it "a great new human adventure," in a speech called "The New Frontier of Space" (1961, 2). Vice President Lyndon Johnson, offering the dedication for yet another NASA spaceflight center, called space the "last and greatest frontier" and, making sure to drive his point home, added "we go into space as pioneers came into the West, for one purpose only" (1963, 1). That goal was more than singular, though, in that the "one purpose" was a metonym for opportunity, historical greatness and, as always, continued freedom. The aspiration of NASA was to achieve lasting historical greatness in a way captured by a historical past believed to be understood by all Americans. NASA flight director Gene Kranz later described his feelings as the Apollo 11 lunar module *Eagle* was seconds from landing on the moon in a fashion that fully encapsulates this mythic *mission*: "There are going to make it! It is like watching Christopher Columbus wade ashore in the New World" (2000, 292).

In a very real mythical sense this version of the frontier offered exactly what Kennedy hoped it would: a driving force for supporting and justifying the support of the Apollo missions. And, as long as government officials and astronauts followed the "pioneering" ideals of that nineteenth-century version of frontier mythology the rhetoric worked effectively as myth, too. But as Gene Kranz's reference to Columbus makes clear it is not the nineteenth-century version of the frontier myth that would be the problem, eventually, for NASA. In referencing William Bradford, Kennedy had offered an earlier version of that frontier myth. That myth was far more closely aligned with providence, a sense of mission, and the emergence of a New World paradise. This version of frontier mythology was, as Charles L. Sanford put it, "an imaginative complex" that presents both images of progress and paradise in an inviting and "dynamic relationship with other values" (1961, vi). It was, in other words, exactly what the astronauts of the 1960s would never find in outer space. And it was one that would, eventually, make Buzz Aldrin's "magnificent desolation" all the more troubling mythically. The reference to Bradford and the Pilgrims in Kennedy's Rice University address acts as a metonymic signifier to a vast "imaginative complex" of a much earlier era, with a much different version of frontier mythology. As a result the NASA missions of the 1960s cannot be simply a continuation of that nineteenth-century trope, but must be considered as the seventeenth- and eighteenth-century mythology evoked by Kennedy in 1961 and 1962.

At the same time, by referencing William Bradford, and through him a host of early colonial writers, Kennedy's Rice address further collapses history again. The Apollo missions are not just a restatement of the nineteenth-century expansionist/progressive history of western movement in the United States but also encompass an earlier sense of the frontier. That earlier sense, which carried with it a specific religious mission, will make specific requirements on the environment of space encountered during the work of NASA in the 1960s. If outer space was to be equated with the wild environment that the first explorers believed they had found on the shores of North America, then space had to be, in Sanford's terms, "the wilderness, which at least in its primitive state suggested Paradise," and when explored by NASA astronauts then, there was required the specific idea that at the very least it could be "convert[ed] into a pleasing land" (1961, 116).

Ironically, then, the Apollo missions were given by Kennedy another, earlier, sense of *mission*—one inherent in an earlier conception of the New World. These interconnected images of the New World as a promised land, a New Jerusalem, and an Earthly paradise, all evoked by the mention of one of the most famous Pilgrim leaders, create for NASA an unachievable mythical mission for their lunar program. Apollo would have to discover, at the very least, some sense of the paradisiacal qualities that Bradford and others saw in New England, Virginia, and indeed the whole of the New World three centuries earlier.

The seventeenth-century sense of mission evoked by Kennedy, then, was an interwoven "imaginative complex" that held together both religious and environmental notions of the New World for the English colonists of the period. Both American history and, later, American Studies would make much use of these complex notions, but in bringing them together in his Rice address, Kennedy presented the lunar missions in a way that those missions could never actually realize. Writing in 1690 the Boston minister Cotton Mather put the mission of Englishmen in New England quite forcefully. The colonists were in the process of creating the perfect church in the New World, and through that work would usher in the Promised Land of a new Eden. Like the Apollo astronauts, the English colonists would have to work to get there, but get there they would. Mather drew a specific image in 1690 that as a Puritan minister concerned with witchcraft he meant to be taken literally but which can also act as a metaphorical statement for NASA in the 1960s: "The Wilderness through which we are passing to the Promised Land is all over fill'd with Fiery flying serpents" (C. Mather and American Imprint Collection [Library of Congress] 1690, 35; original emphasis). For Mather the New World was a place in which English Protestants could achieve the promised biblical paradise. It was to be a place, as the Anglican poet George Herbert offered, in which:

> Religion stands tip-toe in our land
> Ready to pass to the American strand.
>
> (cited in Sanford 1961, 53)

Getting to the New World was arduous, expensive, and dangerous. But once there the promise of a new beginning in a place that was, as Mircea Eliade put it, "*a country where they could be born anew*" dominated the cultural experience(Eliade 1969, 98; original emphasis). The "American strand" of such Edenic beliefs carried a sense of mission for early colonialists was so closely aligned with millennialist notions of paradise that it was the mission of those colonists to restore to Earth the paradise promised by Christ. "Why may not that be the place of the New Jerusalem?" asked an Anglican clergyman around 1625 as he noted "the Gospel's fleeing westward" (cited in Neill 1871, 177). As Cotton Mather's father, Harvard College president and Puritan minister Increase Mather would put it, should the colonists live and worship in the appropriate fashion then "the Church of God shall have a glorious time on Earth," and "this Earth will be restored to its paradise state" (1710, i, 57). This was the promise the English colonists believed the New World offered them—a mission based on biblical prophecy that was obvious, they thought, in the very riches of the land they had "discovered." As John Smith famously claimed of Virginia in 1612 it was a land superior to "the most pleasant places of *Europe*, *Asia*, *Africa*, or *America*," and, indeed, that "heaven and earth never agreed better to frame a place for means of habitation" (1973, 2). In contrast the Apollo astronauts would find a terrain that was barren and lifeless, what Buzz Aldrin would describe as "grey ash-colored lunar scenery, pocked with thousands of little craters and with every variety and shape of rock" (Aldrin 2009, 33). Smith described Virginia with imagery that was far more vibrant and inviting, a scene that was

> not mountanous nor yet low but such pleasant plaine hils and fertle valleyes, one prettily crossing an other, and watered so conveniently with their sweete brookes and christall springs, as if art it selfe had devised them.
>
> (1973, 3)

In Virginia, as in New England and indeed all of the New World for the English, the colonists were presented, according to John Cotton, a "grand charter given to *Adam* and his posterity in Paradise" by undertaking the work to built paradise in the New World (From The Church's Resurrection, cited in Sanford 1961, 85; original emphasis). Whether describing New England, or Virginia, the New World for these English explorers and eventual colonists was a place that "is nothing inferior to Canaan of Israel" as Thomas Morton put it in 1638, "but kind of parallel to it, in all points" (1947, 15).[10] At the same time, when describing this potential Paradise of the New World, John Smith could sound almost like a seventeenth-century Kennedy arguing for Apollo. As Kennedy called for the lunar voyage "not because it is easy, but because it is hard," Smith, describing New England in 1616, noted that to colonies there "requires all the best parts of Art, Judgement, Courage, Honesty, Constancy, Diligence, and Industrie, to doe but neere well" (From A Description of New England [1616] reprinted in Gunn 1994, 97).

More specifically, perhaps, the land of the New World was to be not just paradise but was also, as James Oglethorpe would eventually claim for the colony of Georgia in 1733, a place where "there are fertile lands sufficient to subsist all the useless Poor in England, and distressed Protestants in Europe" (1947, 4). Unlike the that which awaited astronauts in space, or on the surface of the moon, the New World was, as the historian James Horn put it in describing the view of a seventeenth-century Chesapeake Bay, a place where "rivers teamed with fish, forests abounded with game, and skies were darkened with innumerable birds: abundance and plenty" (1994, 128). John Smith, describing the wildlife and environment of Virginia, captures Horn's "abundance" with great detail: "of beasts the chiefe are deare, nothing differing from ours . . . towards the heads of the rivers, ther are many, but amongst the rivers few," he wrote in 1612. "Their squirrels some are neare as greate as our smallest sort of wilde rabbits," he continued, also describing "flying squirrels" that were seen "to fly 30 or 40 yards," while the "Hares" of Virginia were "no bigger than our conies" and wolves "that cannot barke but howle," and were "not much bigger than our English Foxes" (1973, 13–14). For Smith, like Bradford in New England, the chief descriptor of the New World was the sheer abundance of life. The many rivers, the large squirrels, and other wildlife dominate Smith's narrative creating an image of what would come to be considered a land of plenty.

The commander of Apollo 17, Eugene Cernan, found no such abundance at the landing site of Taurus-Littrow in 1972, where there was "not so much as a squirrel track to indicate any sort of life, not a green blade of grass to color the bland, stark beauty, not a cloud overhead, or the slightest hint of a brook or stream," he later recalled of the landscape that greeted him as he descended for his lunar module. And, while Cernan might have "felt comfortable, as if I belonged [there]" it most certainly was not the landscape that John Smith, or Bradford, or the Mather family, or many other English explorers and colonists believed they had found in the New World two or three centuries earlier (1999, 321–322). And while Cernan found space to be metaphorically as "alive and vibrant" during his spacewalk on Gemini 9 as early English explorers found the New World, he was also aware that it was "dead and empty" (1999, 131). Not "fertile lands sufficient to subsist all the useless Poor," the moon was a world far more like the Wyoming described by American philosopher and historian John Fiske in 1887, who described *that* wilderness as "a land of utter desolation, a land where no man could live! It struck me as being like the moon, yes, these terrible mountains, casting their sharp black shadows across the blazing sunshine are the very mountains I have seen through the telescope in the moon!" (John Fiske in a letter to his wife, qtd. in Drinnon 1997, 232).

Kennedy had made his new frontier a frontier that was supposed to work along multiple levels with what he believed Americans saw as their Old Frontier. The moon was to recapture a mythic past, Eliade's "desire to start out again," and offer a virgin world to do it upon. It was always impossible that the lunar surface could ever imply what the mythic rhetoric demanded of it. To be what was necessary for the Apollo missions to truly be a success the moon would have to be, as Edward Williams described Virginia in1650, better than any known nation it could be compared to, with "the same bounty of Summer, the same milde remission of Winter, with a more virgin and unexhausted soyle being materiall arguments to shew that modesty and truth receive no diminution by the comparison" (1947, 11). Instead, it was far more dangerous an environment than the English colonists, for all the deprivation they suffered, ever had to deal with. NASA was fully aware that the environment there had set out to pioneer could never match such paradisiacal images. As Gene Kranz later described the lunar surface, it was a place "with no air, no moisture. The temperatures range from plus to minus 250 degrees Fahrenheit . . . This heavenly body has never seen an earthling" (2000, 277–278). As Command Module Pilot Michael Collins recalled thinking as he waited for Neil Armstrong and Buzz Aldrin to return from the lunar surface as he orbited the moon during Apollo 11, it was a place where if he heard "one little hiccup [then] they were dead men" (2009, 412).

Yet NASA—or at least some of its astronauts—seemed inclined to consider the deeper, more mythical implications of the Apollo voyages. Buzz Aldrin would famously take a Catholic communion before leaving the *Eagle* for his walk on the Sea of Tranquility. Frank Borman, Bill Anders, and Jim Lovell, sixty miles above the moon in orbit aboard Apollo 8, would read the first ten verses of the book of Genesis to an entranced Earth-bound audience on Christmas Eve 1968. Those verses, about the creation of the world, the separation of light from darkness, land from water, and, most important, the heavens from the Earth, suited the ongoing mythical mission of the Apollo program perfectly. Perhaps ironically, the book of Genesis was also cited in relation to the colonies of New England. Likely written by the Reverend John White of Dorset, England, and later found among the papers of Increase Mather in Boston, *The Planters Plea* does not consider the creation of the world as the astronauts of Apollo 8 did. Rather, it is most concerned with that part of Genesis, which demands that the faithful "set them their taske, to replenish the earth, and to subdue it" (White 1947, 1). The astronauts of Apollo 8 could offer no similar task, however. The lunar surface was, as Buzz Aldrin would later declare, nothing more than magnificent desolation. But along with their reading of Genesis the three astronauts of Apollo 8 would also offer their impressions of the lunar surface to a world audience during their lunar orbits. Their description would, in a very real way, capture exactly why the task set out by Kennedy at Rice University, adopted by NASA and the Apollo Program, and brought to fruition in

July 1969 would never be the mythically resonant event that it was implied to be. The astronauts could not "go forth and subdue" the lunar surface, after all. It would not be, as John Smith declared of New England, a place where there were "all sorts of fish . . . [and] in a land most fertile, temperate, and plentifull of all necessaries for the building of ships, boats and houses, and the nourishment of man" (Smith 1947, 9).

Apollo 8 saw none of the required "plentifull necessaries" on the surface of the moon. Jim Lovell, the command module pilot who would later be the commander of the failed Apollo 13 mission, remarked, "The vast loneliness up here of the moon is awe inspiring, and it makes you realize just what you have back there on earth. The earth from here is a grand oasis in the big vastness of space." But it was the Mission Commander Frank Borman who offered the most damning of opinions, and one that perfectly articulates why the Apollo missions, mythologically at least, were destined to failure. "I know my own impression is that it's a vast, lonely, forbidding-type existence, or expanse of nothing," he declared. "It certainly would not appear to be a very inviting place to live or work" (NASA Historical Reference Collection 1968, tape 57, 5–6; Lovell and Kluger 1995, 51–54).[11]

Apparently fully convinced of his opinion, Apollo 8 would be Borman's final mission as an astronaut. Where the political interpretations of NASA in the 1960s ends with the successful mission of Apollo 11 in the Cold War–inspired Space Race, and other histories of the period offer the institutional and personal successes of each completed mission from the early Mercury launches to the final use of the Saturn V rockets for Skylab in the 1970s, Borman's opinion must be considered as a different assessment of the period. In John Kennedy's Rice University address, the president had tethered a vast and complex set of mythic images onto his hoped-for space program. But with those mythic images came mythic burdens—and they remained burdens the lunar surface could not fulfill. Long before Kennedy declared his desire for, "before this decade is out," landing men from Earth on the moon, the sense of inevitability of such missions dominated thought about the future of space flight. From the science fiction of generations earlier to the *Collier's* essays declaring "man will conquer space, soon," a segment of the American population believed that, one day, outer space would be the domain of human explorers. John F. Kennedy provided the mythological structure to support such a mission, exactly as American aerospace corporations and NASA provided the technological and organizational structure. And after the moon landing of July 1969 NASA had fully embraced Kennedy's seventeenth-century terms for the missions, answering the question, "But why?" about the Apollo missions by declaring that

> [i]t is an old question, one that must have been raised when the first European settlers crossed the perilous North Atlantic. Why did settlers commit themselves to a harsh and hazardous voyage, to reach at best a howling wilderness of wild animals and savages, with brutal winters in

which they might starve? . . . What possible value could be derived from the American wilderness?

(NASA 1970, 169–170)

Gone was the nineteenth-century frontier rhetoric, replaced by far earlier versions. NASA's mission was never able to present Marx's "new symbols of possibility" or Doty's "forms of the future." Rather it offered conspicuously "old symbols" as it attempted to turn a traditional mythos into something launched by the logos of a Saturn V rocket. Thus, that "But why?" question could not be fully answered mythically by the astronauts of Apollo. Politically, and personally for those involved, the NASA missions of the 1960s were unquestionably a success. Buzz Alrin, who noted he had gone "where no man had gone before" still called his most "significant voyage" his "return" to Earth (Aldrin and Warga 1973, 338). Aldrin seemed to know that mythically NASA failed, and the loss of interest among Americans after the landing of Apollo 11 illustrates this mythic failure as much as its political and technological success. There had been a hope, as Mercury astronaut Scott Carpenter had put it in 1962, that there was "no doubt about it, space is a fabulous frontier, and we are going to solve some of its secrets and bring back many of its riches in our lifetime" (Carpenter et al. 1990, 346). Yet, that sense of inevitability was never fully lost even if the lunar missions could not carry the sense of mission implied by the likes of William Bradford or John Smith. It merely found a new outlet. Writing for president-elect Kennedy in early 1961 as his science advisor, Jerome Wiesner had offered a perspective on that new outlet long before Kennedy was interested in a lunar mission. Wiesner perfectly captured the sense of inevitability of space exploration and, unwittingly, captured as well how popular culture would present that inevitability in ways about which NASA could only dream. Once again answering the "But why?" question and writing about the eventual exploration of space he wrote, "Given his enormous curiosity about the universe in which he lives and his compelling urge to *go where no one has ever been before*, this will be done" (Wiesner 1961, 2; emphasis added).

Of course, "where no man has gone before" would be, mythically as well as practically, the mission of NASA and Apollo. And while Wiesner did not know it yet, it was also the mission of *Star Trek*.

NOTES

1 Logsdon and Kennedy advisor Theodore Sorensen have differing perspectives on this meeting. See, also, Sorensen (2008, 334–336).

2 Logsdon draws these quotes from the Sorenson files at the Kennedy Presidential Library, including the specific memo cited here, but they also correspond to other, less specific, reports of the same meeting.

3 Sidey's story here is drawn from multiple sources, carefully examined by John M. Logsdon. But see especially Logsdon (1970, 106–108; 2010, 76–78; but see

especially Sidey [1963, 100–103]. Sidey's work, both as a reporter and later in his book on Kennedy, is carefully humble in including his own influence in the stories he covered, however.)

4 These are but two of a large number of committees undertaken in Congress immediately after the launching of Sputnik, including both standing committees and special committees. It is from these committees that the draft bill that would form NASA emerged.

5 Each of the listed sources provides useful histories of the organization of NASA in the period under question. Emme was the official NASA historian from 1960 to 1968, and his work, though underused, remains an excellent source.

6 Besides the *Collier's* essays, themselves, and the work of Heppenheimer already cited, McCurdy (1997, 36–46) also offers details on both von Braun's influence on the original essays and the effect of those essays on eventual NASA planning.

7 The work was used as a major source for a twelve-part HBO miniseries ("From the Earth to the Moon [Miniseries]" 1998) as well.

8 Citations to this speech refer to the paragraph number within the text at the Kennedy Presidential Library. The speech, consisting of thirty-two paragraphs, ran roughly eighteen minutes in length when Kennedy delivered it.

9 Eliade is discussing various Brahamaic myths for the purpose of elucidating a general theme. While those specific myths vary from what is discussed here, the themes are consistent.

10 All sources cited from the *Tracts and Other Papers* are cited by volume and then paginated in the original publication format within each volume.

11 The NASA recordings offer no analysis while Lovell's narrative of the mission provides an insightful emphasis on the reading of Genesis by all three astronauts.

WORKS CITED

Aldrin, Edwin E. "Buzz." 2009. *Magnificent Desolation: The Long Journey Home from the Moon*. 1st ed. New York: Harmony Books.

Aldrin, Edwin E. "Buzz," and Wayne Warga. 1973. *Return to Earth*. 1st ed. New York: Random House.

Anderson, Dwight G. 1988. "Power, Rhetoric, and the State: A Theory of Presidential Legitimacy." *The Review of Politics* 50 (2): 198–214.

Beschloss, Michael R. 1997. "Kennedy and the Decision to Go to the Moon." In *Spaceflight and the Myth of Presidential Leadership*, edited by Roger D. Launius and Howard E. McCurdy, 51–67. Urbana: University of Illinois Press.

Bradford, William. 1908. *Bradford's History of Plymouth Plantation*. Edited by William T. Davis. New York: Charles Scribner's Sons.

Byrnes, Mark E. 1994. *Politics and Space: Image Making by NASA*. Westport, CT: Praeger.

Carpenter, M. Scott, Gordon L. Cooper, John H. Glenn, Virgil I. Grisson, Walter M. Schirra, and Alan B Shepard. 1990. *We Seven*. New York: Simon and Schuster Pbks.

Cernan, Eugene. 1999. *The Last Man on the Moon: Astronaut Eugene Cernan and America's Race in Space*. 1st ed. New York: St. Martin's Press.

Chaikin, Andrew. 2007. *A Man on the Moon: The Voyages of the Apollo Astronauts*. New York, NY: Penguin Books.

Collier's Editors, Wernher Von Braun, Willy Ley, Fred L. Whipple, and Cornelius Ryan. 1952. "Collier's Magazine Man Will Conquer Space Soon." *Collier's Magazine*, October 22.

Collins, Michael. 2009. *Carrying the Fire: An Astronaut's Journeys*. New York: Farrar, Straus, and Giroux.

Cooper, Jr., Henry S. F. 1969. *Apollo on the Moon*. New York: Dial Press.

De Groot, Gerard J. 2006. *Dark Side of the Moon: The Magnificent Madness of the American Lunar Quest*. New York: New York University Press.

Doty, William G. 1980. "Mythophiles' Dyscrasia: A Comprehensive Definition of Myth." *Journal of the American Academy of Religion* XLVIII (4): 531–62.

———. 1995. "Silent Myths Singing in the Blood: The Sites of Production and Consumption of Myths in a 'Mythless' Society." In *Picturing Cultural Values in Postmodern America*, edited by William G. Doty, 187–220. Tuscaloosa: University of Alabama Press.

———. 2000. *Mythography: The Study of Myths and Rituals*. 2nd ed. Tuscaloosa: University of Alabama Press.

Drinnon, Richard. 1997. *Facing West: The Metaphysics of Indian-Hating and Empire-Building*. Norman, OK and London: Oklahoma University Pres.

Eliade, Mircea. 1954. *The Myth of the Eternal Return*. New York: Pantheon Book.

———. 1969. *The Quest: History and Meaning in Religion*. Chicago: University of Chicago Press.

Emme, Eugene Morlock. 1966. *A History of Space Flight*. Holt Library of Science. New York: Holt, Rinehart and Winsto.

Felkins, Patricia K., and Irvin Goldman. 1993. "Political Myth as Subjective Narrative: Some Interpretations and Understandings of John F. Kennedy." *Political Psychology* 14 (3): 447–67.

From the Earth to the Moon (Miniseries). 1998. Los Angeles: HBO.

Fulmer, Hal W. 1986. "A Rhetoric of Oldspeak: Mythic Elements in Presidential Inaugural Addresses, 1960–1980." *Rhetoric Society Quarterly* 16 (4): 299–312.

Gallup, George Horace. 1972. *The Gallup Poll: Public Opinion 1935–1971*. New York: Random Hous.

Glenn, John H. 1961. "We're Going Places No One Has Ever Travelled in a Craft No One's Flown." *Life*, January.

Grey, Jerry, and Vivian Grey. 1962. *Space Flight Report to the Nation*. New York: Basic Book.

Gunn, Giles B., ed. 1994. *Early American Writing*. New York: Penguin Books.

Heppenheimer, T. A. 1997. *Countdown: A History of Space Flight*. New York: John Wiley & Sons.

———. 1999. *The Space Shuttle Decision: NASA's Search for a Reusable Space Vehicle*. NASA SP 4221. Washington, DC: National Aeronautics and Space Administration, NASA History Office, Office of Policy and Plans.

Holmes, Jay. 1962. *America on the Moon: The Enterprise of the Sixties*. 1st ed. Philadelphia: Lippincot.

Horn, James P. 1994. *Adapting to a New World: English Society in the Seventeenth-Century Chesapeake*. Chapel Hill: Published for the Institute of Early American History and Culture, Williamsburg, Virginia, by the University of North Carolina Press.

Johnson, John. 1961. "The New Frontier of Space." Public Speech presented at the New York Patent Law Association, NASA History Office, March 24.

Johnson, Lyndon Baines. 1963. "Vice President Lyndon Johnson, Space Center Dedication." NASA History Office, November 14.

Jordan, John W. 2003. "Kennedy's Romantic Moon and Its Rhetorical Legacy for Space Exploration." *Rhetoric & Public Affairs* 6 (2): 209–32.

Kauffman, James Lee. 1994. *Selling Outer Space: Kennedy, the Media, and Funding for Project Apollo, 1961–1963*. Tuscaloosa: University of Alabama Press.

Kennedy, John F. 1960. "Address of Senator John F. Kennedy Accepting the Democratic Party Nomination for the Presidency of the United States." Presented at the Democratic National Convention, Memorial Coliseum, Los Angeles, July 15.

———. 1961. "Special Message to the Congress on Urgent National Needs." United States Congress, May 25.

———. 1962. "Address at Rice University on the Nation's Space Effort." Rice University, Texas, September 12.

Kranz, Gene. 2000. *Failure Is Not an Option: Mission Control from Mercury to Apollo 13 and Beyond*. New York: Simon & Schuster.

Launius, Roger D. 2004. *Frontiers of Space Exploration*. Westport, CT: Greenwood Press.

Life Magazine Editors. 1961. "A Costly Trip to the Moon." *Life*, June 16.

Logsdon, John M. 1970. *The Decision to Go to the Moon: Project Apollo and the National Interest*. Cambridge: MIT Press.

———. 2010. *John F. Kennedy and the Race to the Moon*. New York: Palgrave Macmillan.

Lovell, Jim, and Jeffrey Kluger. 1995. *Lost Moon: Apollo 13*. New York: Pocket Book.

Low, George M. 1960. "Project Mercury Progress." Presented at the UPI Editors Conference, NASA History Office: Washington, DC, December 9.

Martin, Ralph G. 1983. *A Hero for Our Time: An Intimate Story of the Kennedy Years*. New York: Macmillan.

Mather, Cotton, and American Imprint Collection (Library of Congress). 1690. *The Present State of New-England. Considered in a Discourse on the Necessities and Advantages of a Public Spirit in Every Man; Especially, at Such a Time as This. Made at the Lecture in Boston 20. D. 1. M. 1690. Upon the News of an Invasion by Bloody Indians and French-Men, Begun Upon Us*. Boston: Printed by S. Green.

Mather, Increase. 1710. *A Discourse Concerning Faith and Fervency in Prayer, and the Glorious Kingdom of the Lord Jesus Christ, on Earth, Now Approaching*. Boston: T. Green.

McCurdy, Howard E. 1997. *Space and the American Imagination*. Washington, DC: Smithsonian Institution Press.

McDougall, Walter A. 1985. *The Heavens and the Earth: A Political History of the Space Age*. New York: Basic Books.

Morton, Thomas. 1947. "New English Canaan, or New Canaan." In *Tracts and Other Papers, Relating Principally to the Origin, Settlement, and Progress of the Colonies of North America, From the Discovery of the Country to the Year 1776*. Vol. 2. New York: Peter Smith.

NASA. 1959. *The Challenge of Space Exploration*. Washington, DC: Government Printing Office.

———. 1960. *House Committee on Science and Astronautics*. Washington, DC: United States Congress: Government Printing Office.

———. 1965. *Space Exploration—Why and How (NASA EP-25)*. Washington, DC: Government Printing Office.

———. 1970. *This Island Earth (NASA SP-250)*. Edited by Oran W. Nicks. Washington, DC: U.S. Government Printing Office.

NASA, and David A. Anderton. 1969. "Man in Space (NASA EP-57)." Washington, DC: Government Printing Office.

NASA Historical Reference Collection. 1968. "Apollo 8 Technical Air-to-Ground Voice Transmission." NASA.

———. 1969. "Apollo 11 Technical Air-to-Ground Voice Transmission." NASA.

National Aeronautics and Space Act of 1958 (as Amended). 1958. 72 *Stat*. 426. Vol. USC 2451 et seq.

National Aeronautics and Space Administration, and NASA Office of Program Planning and Evaluation. 1959. *NASA Long Range Plan*. United States Government Report. Washington, DC: NASA Historical Reference Collection, History Office.

Neill, Edward D. 1871. *The English Colonization of America*. London: Strahan & Co.

New York Times Editors. 1961. "Editorial." *New York Times*, May 26, sec. A.

Oglethorpe, James. 1947. "A Brief Account of the Establishment of the Colony of Georgia under General James Oglethorpe." In *Tracts and Other Papers, Relating Principally to the Origin, Settlement, and Progress of the Colonies of North America, from the Discovery of the Country to the Year 1776*, Vol. 1, 1–15. New York: Peter Smith.

Rushing, Janice Hocker. 1986. "Mythic Evolution of 'The New Frontier' in Mass Mediated Rhetoric." *Critical Studies in Mass Communication* 3: 265–96.

Sanford, Charles L. 1961. *The Quest for Paradise: Europe and the American Moral Imagination*. Urbana: University of Illinois Press.

Schefter, James L. 1999. *The Race: The Uncensored Story of How America Beat Russia to the Moon*. 1st ed. New York: Doubleday.

Sheehan, Michael. 2007. *The International Politics of Space*. London: Routledge.

Shelton, William Roy. 1967. *American Space Exploration: The First Decade*. Boston: Little, Brown.

Sidey, Hugh. 1961. "The Week's Events: How the News Hit Washington." *Life*, April 21.

———. 1963. *John F. Kennedy, Portrait of a President*. 1st ed. New York: Atheneum.

Smith, Henry Nash. 1971. *Virgin Land: The American West as Symbol and Myth*. Cambridge, MA: Harvard University Press.

Smith, John. 1947. "New Englands Trials, Declaring the Successe of 80 Ships Employed Thither within These Eight Years; and the Benefit of That Countrey by Sea and Land." In *Tracts and Other Papers, Relating Principally to the Origin, Settlement, and Progress of the Colonies of North America, From the Discovery of the Country to the Year 1776*. Vol. 2. New York: Peter Smith.

Smith, John. 1973. *A Map of Virginia with a Description of The Coventrey, The Commodities, People, Government and Religion*. Amsterdam: Theatrum Orbis Terrarum Ltd.

Sorensen, Theodore C. 2008. *Counselor: A Life at the Edge of History*. 1st ed. New York, NY: Harper.

Turner, Frederick Jackson. 1999. "The Significance of the Frontier in American History." In *Does the Frontier Experience Make America Exceptional?*, edited by Richard W. Etulain, 17–43. Boston: Bedford/St. Martin's.

U.S. Congress, House of Representatives. 1958. Journal of the House of Representatives of the United States, 85th Congress, 2d Session.

"Von Braun Testimony on Space Plans." 1963. *Congressional Record* 19250 (October).

Watanabe, Hirotaka. 2009. "The Kennedy Administration and Project Apollo: International Competition and Cooperation through Space Policy." *Osaka University Law Review* 56: 31–48.

White, John (Rev.). 1947. "The Planters Plea, Or, The Groves of Plantations Examined, and Vsuall Objections Answered." In *Tracts and Other Papers, Relating Principally to the Origin, Settlement, and Progress of the Colonies of North America, From the Discovery of the Country to the Year 1776*. Vol. 2. New York: Peter Smith.

Wiesner, Jerome B. 1961. *Report to the President-Elect of the Ad Hoc Committee on Space (Wiesner Committee)*. Washington, DC: NASA Historical Reference Collection, History Office.

Williams, Edward. 1947. "Virginia, More Especially the South Part Thereof, Richly and Truly Valued Viz. the Fertile Carolana, and No Lesse Excellent Isle of Roanoak, of Latitude from 31 to 37 Degr. Relating the Meanes of Raysing Infinite Profits to the Adventurers and Planters." In *Tracts and Other Papers, Relating Principally to the Origin, Settlement, and Progress of the Colonies of North America, from the Discovery of the Country to the Year 1776*. Vol. 3. New York: Peter Smith.

5 The Rejection of Paradise
Star Trek and the Final Frontier

Space, The Final Frontier . . .

—Opening litany of *Star Trek*, 1966–[1]

James T. Kirk was the youngest captain in Starfleet history when he began his five-year mission aboard the United Federation of Planets' starship, the USS *Enterprise*.[2]

With a mostly human crew, a half-human, half-Vulcan first officer named Spock, and a futuristic military-industrial complex fully at his beck and call, Kirk would become one of the most successful captains in Starfleet history. It is likely significant that Captain James T. Kirk was also wholly a fictional character, created by Gene Roddenberry and the producers of *Star Trek* over the course of three television seasons from 1966–1969 and expanded significantly later. Specifically created as a "Captain Horatio Hornblower" of the twenty-third century according the *Star Trek* producers in their "writer's guide" for the second season of the show, the character of Kirk had a "broad and highly mature perspective on command, fellow crewmen, and even on alien life customs" (Star Trek Producers 1967, 10).

This is a positive statement of the mission of Captain Kirk and the *Enterprise* but not an entirely foreign one to American viewers in the 1960s; one of Captain James Kirk's possible namesakes, James Kirke Pauling, had similarly described such qualities in his 1832 book *Westward Ho!* That work seemed to describe the mission of the *Enterprise*; for Kirke Pauling, Captain Kirk's true mission might have been one that while going "where no man has gone before" it was also a given that, "no matter where he goes, to whatever region . . . [he] carries with him his destiny, which is to civilize the world, and rule it afterwards" (Pauling 1832, 1:94). Such historical precedents are easily traceable for Kirk—creator Gene Roddenberry had claimed he sold the show to the National Broadcasting Company (NBC) as "a *Wagon Train* to the stars" after all (Whitfield and Roddenberry 1968, 23). Yet fans of Kirk saw him in a strictly positive light. Early *Star Trek* fans and convention organizers Jacqueline Lichtenberg, Sondra Marshak, and Joan Winston described Kirk and his fans this way:

> [F]or Kirk fans . . . there has never been another hero quite like him . . . Kirk is the Ideal Man. They see a Kirk who is unafraid of his emotions yet very definitely masculine. He is aggressive without being brutal. He is compassionate without being weak. He is decisive without being officious. He is efficient without being rigid.
>
> (1975, 39; original emphasis)

One of the writers who contributed to the original series, David Gerrold, put this description somewhat more specifically. James T. Kirk had to be an "explorer, ambassador, soldier, and peacekeeper. He would be the sole arbiter of Federation law wherever he travelled—he would be a law unto himself" (Gerrold 1973, 6). And, as a "law unto himself" Kirk's mission was as much to spread the moral certitudes of *Star Trek*'s United Federation of Planets to the rest of the galaxy as much as it was a mission of discovery and exploration—perhaps even more so. Philosopher John Shelton Lawrence and theology scholar Robert Jewett explain the narrative assumptions of the show more critically, claiming that "*Star Trek*'s innocent optimism conceals the unexamined premise that the 'American way of life' will somehow prevail in the universe" (2002, 257). Indeed, it is a joke among fans of the show that Kirk repeatedly breaks the most important rule claimed by Federation law almost at whim, always to promote what Lawrence and Jewett call that "American way of life." That rule, originally called "General Order Number One" and later, simply, "the Prime Directive," requires that Federation explorers like Kirk not interfere with the "natural development" of the alien cultures they encounter. As American film scholar Katrina G. Boyd has described it, however, it is this rule that contains the main contradiction in the logic of *Star Trek*:

> The "Prime Directive," which restricts the Federation and its Starfleet officers from interfering in the "natural" development of another culture, not only implies noninterference but also confirms the law of progress. The Prime Directive enacts the show's fundamental conflict between a belief in the exercise of free will and self-determination and the existence of some necessary, natural pattern of development, which "common sense" tells us involves unending technological and moral progress.
>
> (1996, 102)[3]

When Kirk disregards the Prime Directive it is always in favor of the "natural development" that will put a local culture on the road to "unending technological and moral progress." Gerrold has put it even more succinctly, claiming that the *Enterprise* is actually "meddling her was across the galaxy . . . to spread truth, justice, and the American Way to the far corners of the universe." The mission of Captain Kirk is confirmed, by Gerrold, when he notes that "if a local culture is tested and found wanting in the eyes of a starship captain, he may make such changes as he feels necessary." And, in *Star Trek*, those changes are often radical, often painful to the local culture, and always place them on a road toward "progress," "because," Gerrold notes, "if our ends are just, then no matter what means we choose, our means will be just, too" (1973, 256; original emphasis).

Thus, although fictional, Captain Kirk has become, in the decades since the original airing of the show, perhaps one of the most well-known fictional American characters created in the 1960s. He is, as John William Ward wrote of Andrew Jackson, a "creation of his time . . . of [which] the people made a mirror for themselves" (1955, 208). So important is Kirk that perhaps his fictional status matters little. As Lichtenberg, Marshak, and Winston put it, as far as any fan—and perhaps American culture—is concerned "Kirk *is* real" (1975, 41; original emphasis).

Indeed, characters—whether real or imagined—have been central to frontier mythology for as long as that mythology has existed. In *The Western Hero in History and Legend* Kent Ladd Steckmesser examined a similar character—although, in this case, at least partially an actual historical individual. That person, "Wild Bill" Hickok, was a centrally important part of the nineteenth-century frontier myths in the United States. Hickok was to have confronted an allegedly train-robbing group, known as the "McCanles Gang," and participated in a gunfight against them in which Hickok supposedly killed some members of the gang. Steckmesser's research found that "by 1931" the gunfight with the McCanles Gang and Hickok "had appeared in an autobiography, (1907), a novel (1923), a movie (1923), a *Saturday Evening Post* article (1926), and countless newspapers" (Steckmesser 1997, 247; Kapell 2010a, 214–215).[4]

The literature on Captain Kirk is easily larger than that Steckmesser detailed for Hickok. As I have noted elsewhere,

> By 2009 it is possible to trace the character through the original *Star Trek* (1966–9) series, seven films (1979–1994), and William Shatner's autobiographies of his time on the original series: *Star Trek Memories* (1993), of the films, *Star Trek Movie Memories* (1994) and his quasi ethnographic examination of his fellow actors and *Trek* fans, *Get A Life*! (1999). William Shatner has even added to the mythos of James T. Kirk by co-writing many of his own . . . *Trek* novels, including *The Return* (1996) . . . There are countless other novels such as Diane Carey's story of the young Jim Kirk's second trip into space, *Best Destiny* (1992), and multiple award-winning writer Vonda McIntyre's version of the very first mission of the entire crew: *Enterprise: The First Adventure* (1986). Add to this countless contributions of fan fiction from the late 1960s onward, including "slash" fiction positing a romantic relationship between Kirk and Spock . . . [and] there is also the fan produced "second five-year mission" of the *Enterprise* in *Star Trek Phase*

> *II* (2003-present) with James Cawley as Kirk, and the newest *Star Trek* (2009) with Chris Pine as Kirk. In short, there is as much material for James T. Kirk as there ever was for "Wild Bill" Hickok.
>
> (Kapell 2010a, 215)

Comparing Captain Kirk to the likes of a "Wild Bill" Hickok leads to a relatively obvious conclusion mythologically. "It is not much of a stretch," American anthropologist Jon Wagner and health sociologist Jan Lundeen have noted, "to view Captain Kirk as *Star Trek*'s answer to the classic Western hero." On board the *Enterprise* Kirk represents that archetype "upholding orderly society but never quite contained by it" (Wagner and Lundeen 1998, 177). The *Americaness* of this quality is, ironically, self-evident. As American classicist Wm. Blake Tyrrell noted in 1977, "the western story is *the only indigenous mythic narrative of the white American*," and Kirk represents that story and that archetype expanded to an interstellar scope (2010b, 19; emphasis added).[5] In doing so *Star Trek* offered—and continues to offer—viewers "our mythological roots disguised as branches" and presents "American myths clothed in the garb of science fiction" (Tyrrell 2010a, 28, 2010b, 20). As noted when discussing Joe Haldeman in chapter three, this is the theme offered by David Mogen, that "the frontier continues to be the place where American ideals are defined, where American heroes are reborn," and Kirk represents precisely this process (1993, 15). While this is, actually, a major trope of *all* American science fiction, it is one that is employed quite intensely in *Star Trek*. As the American urban studies specialist Carl Abbott has claimed using *Star Trek* as a major example, "much of science fiction is a search for places to start over, for new 'geographies of hope' " (2006, 14). Within the American mythscape, such a notion long predates the emergence of science fiction, however. Eliade's claim that for centuries "both the first colonist and the later European immigrants journeyed to America as *the country where they might be born anew*, that is, begin a new life," has a special relevance within the narratives of *Star Trek* (1969, 98; original emphasis). The show's five-year mission had Kirk leading his crew, time and time again, to planets defined by their "newness." The American mythic desire to remake the world anew offered, as I have previously noted elsewhere, a "perfectly reasonable" justification for Kirk and his crew to be "forever interfering with other cultures they encounter, even though such behavior is against their own Prime Directive" of noninterference (Kapell 2010c, 69–70).

But while Kirk occupied a central, archetypal role in the original *Star Trek* series, the other characters have also entered the popular culture canon. Science Officer Spock and his declarations of "illogical," or Dr. McCoy's "Are you out of your Vulcan mind?" have made the characters, fictional planets, and the starship *Enterprise* immediately recognizable images for the forty-five years since the original episodes began airing on NBC in 1966. As American scholar Richard Thomas describes it, with "words like 'phaser' and 'warp,' phrases like 'beam me up' and 'space, the final frontier,'" *Star Trek* has become "part of the vocabulary of modern culture" (1997, 2). David Gerrold can go so far as to proclaim it "a genre unto itself," and, for its owners, Paramount Studios, *Star Trek* has become simply called "the franchise" (1973, 15; Westfahl 1996). And as a franchise, its expansion has followed a similar journey of that of an earlier generation's "westerns," which achieved cultural omnipresence through films, television, inexpensive "dime novels," merchandise, and a multitude of other cultural products. As I have noted,

> after the first *Star Trek* convention in 1972 [the franchise became] a cultural construct unlike any other. The folklore of *Trek* conventions, the contemporary "dime novel" manifestations of the hundreds of inexpensive mass-market paperback novels, the fan fiction written by uncounted numbers of individuals and groups, all contribute to the *mythos* of *Star Trek*.
>
> (Kapell 2010a, 214; original emphasis)

That "cultural construct unlike any other" has produced eleven feature films, five television series—six if one included the animated series of 1973–1974—and, as of 2009, 750 novels currently in print (Kapell 2010b, 2). This multitude of media products almost all present *Star Trek* in the same fashion as well. As American film and literature scholar Ace G. Pilkington described it, "[m]ore than anything else *Star Trek* took the Western's sense of freedom, of unlimited horizons and untrammelled action," and moved it to the "final frontier" of space (2010, 57). At the same time scholarly attention to *Star Trek* has also become, much like the franchise itself, a veritable industry with almost "unlimited horizons" of critical perspectives. And, because of the wealth of both source texts and critical examinations of that material, no contemporary franchise in popular culture has been as closely examined as *Star Trek*. While political scientists and historians have largely examined the other topics of this work, *Star Trek* has been the purview of cultural studies scholars. The troika of "race, gender and class" has been the most popular of topics as a result (During 2005). *Trek*'s racist stereotypes, which Daniel Leonard Bernardi points out in detail as having presented "the integrated supporting cast . . . at the margins of most stories and in the background of most shots," has been of concern to many (1998, 39).[6] Elyce Rae Helford, in her discussion of Captain Kirk's masculinity, could also note "that the original *Star Trek* . . . is sexist hardly needs articulation" (1996, 11).[7] And, for most, both gender and race, as common descriptors of "the Other" have operated in *Star Trek* to produce narratives that resonate with discussions of class as well. It is important to note that while looking back at *Star Trek* with contemporary eyes leads to images of racism, sexism, imperialism, and a host of other "isms," unlike other television shows of the period *Star Trek* was, as Australian media scholar Bruce Isaacs has put it, "at least . . . willing to engage dialectically with issues of race, gender and sexuality, national and foreign policy, [and] notions of Self and Other" (2010, 184). Throughout the course of its initial run, and through the many spin-offs produced by the franchise, *Star Trek* has provided, as Isaacs describes it, extremely consistent "philosophical foundations" that, in its influence, far outstrips any excessive academic critiques (Isaacs 2010, 185).

In part this abundance of critical perspectives are possible because *Star Trek*, as science fiction, is open to such interpretations and critiques. This is a singular value of science fiction, as a genre, because it operates outside the boundaries of a normal, empirical reality. American literary scholar Carl D. Malmgren, in describing this aspect of science fiction, has noted that a main purpose of the genre is to provide the separation from reality primarily to offer a divergent perspective *on* that reality. In Malmgren's terms, *Star Trek* "distances the [viewer] from the empirical world, generating a cognitive space which the [viewer] must negotiate . . . a science fiction world is thus less a reflection *of* than a reflection *on* empirical reality" (1991, 11; original emphasis).[8] That is, *Star Trek* should be considered a television series that is as much "about the 1960s" as it is merely a product of that period. David Gerrold, writing in 1973, explained this succinctly, offering that the "crew of the *Enterprise* is twentieth century America in space . . . They are representative of the American Sphere of Influence *today*" (47; original emphasis). And, because the 1960s were themselves such a tumultuous period, the differing perspectives and analytical positions that can be used in examining *Star Trek* are, themselves, radically diverse.

Yet, regardless of how scholars have approached *Star Trek* all have been forced to begin, in some fashion or other, those "philosophical foundations." And those foundations begin, as did each episode of the series, with the famous opening narration. At the beginning of all but two episodes viewers heard the following:

> Space: the final frontier. These are the voyages of the starship *Enterprise*. Its five-year mission: to explore strange new worlds, to seek out new life and new civilizations, to boldly go where no man has gone before.[9]

There is no way to read this narration, Valarie Fulton has suggested, as anything other than classic imperialism. She notes of *Star Trek*'s governing body, the Federation, that its "goals are both 'to seek out new civilizations' and 'to boldly go where no one has gone before'—missions that clearly contradict each other unless read through the lens of frontier ideology, which grants new civilizations existence only to the extent that the originary culture has 'found' them" (Fulton 1994, 6). This is not a cultural paradigm unique to *Star Trek*, however. Jay Goulding, a scholar of Japanese and Chinese sociology who set aside his academic specialty long enough to note in his *Empire, Aliens and Conquest* that "the frontier spirit, the quest for knowledge, identification with heroes, the sweet taste of victory in battle are all fundamental themes to *Star Trek*—and to western democracies" (1985, vii). Many have noted this connection of *Star Trek*'s "frontier" themes as part of their larger critiques, such as Gregory M. Pfitzer, who declares that "*Star Trek* continued to celebrate the space cowboy" and that as such "appealed to something inherently mythical and western in viewers." At the same time, for Pfitzer, it was part of a larger tendency of American science fiction to point to alien-hating as a stand in for the "metaphysics of Indian-hating on the western frontier" (1995, 55, 64).

At the same time, however, James T. Kirk and the crew of the starship *Enterprise* has also been seen in vastly more positive ways. Where many scholars have found examples of cultural imperialism in *Star Trek*'s frontier mythology, others declared it racist or sexist. American anthropologist Conrad Phillip Kottak has offered a far more generous analysis. For Kottak *Star Trek*'s narrative was about the "meeting of the 'other'" that is very much a genre theme in science fiction. So where others see the marginalization of characters of color and aliens, Kottak sees a science fictional representation of the first Thanksgiving described by William Bradford in his *History of Plymouth Plantation*. Thus, for Kottak, the very crew of the *Enterprise* should be read through a more optimistic and eventually assimilationist lens because "The Starship *Enterprise* is a melting pot" and "*Star Trek* proclaims [the] sacred principles that validate American society." He takes this analysis even further, proclaiming that

> Kirk is . . . symbolic of the original Anglo-American . . . McCoy's Irish (or at least Gaelic) name represents the next wave, the by now established immigrant. Sulu is the successfully assimilated Asian-American. The African-American female character, Uhura, whose name means "freedom," proclaims that blacks will eventually become full partners with all other Americans.
>
> (Kottak 1990, 103)

Kottak finds in the Russian Ensign Pavel Chekov and the half-human Mr. Spock—and indeed the entire crew of the *Enterprise*—a narrative about "convincing and assimilating, rather than conquering" (Kottak 1990, 105). He sees in *Star Trek*, then, a very optimistic form of American mythology, but mythology nonetheless.

THE ANTIMYTHIC BIAS OF *STAR TREK*

By now it is a relatively standard academic position to note that all of *Star Trek* acts mythologically in contemporary society. Beginning in 1977 when Wm. Blake Tyrrell published his seminal essay, "*Star Trek* as Myth and Television as Mythmaker," a significant number of articles and books have been published contending that the franchise should be read mythically (Kapell, 2010a). In both Thomas Richards's *Star Trek in Myth and Legend* and my own volume, *Star Trek as Myth*, the contention has been made that *Star Trek* works by "retelling stories as old as our species in ways understandable to the modern and postmodern viewer" (Kapell 2010a, 217; also Richards 1997; Kapell 2010d). What is surprising about the sheer number of works offering mythographic readings of the franchise is that *Star Trek*, from the very beginning, has offered itself as a more realistic form of science fiction—or, frankly, as a nonfantasy, nonmythic narrative. The writers and producers of *Star Trek* worked hard to present a plausible show, after all. As Jon Wagner describes it, "*Star Trek* bears the imprint of a stridently rationalist humanism" that distinguishes it from other science fiction and fantasy works of popular culture (1999, 119). Stephen E. Whitfield put this somewhat more specifically in his *The Making of Star Trek*: the show "was a combination of speculation, inspiration, and (within the limits of practicality) verification with scientific sources. Literally thousands of questions were asked, covering even the tiniest of details" (Whitfield and Roddenberry 1968, 163).

To be doubly sure that *Star Trek* was read as science fiction rather than space fantasy the show's producers offered specific episodes that were designed to fully undermine myths in general. Most famously, the second-season episode "Who Mourns for Adonais?" offered a specific attack on myth with the strongly implied message that, in the twenty-third century, myth would be seen as counterproductive and something that humanity had clearly abandoned (Daniels 1967). In the episode the *Enterprise*, en route to a habitable planet, confronts a giant hand in space that imprisons the ship. A laurel wreath-wearing man's face then appears on the ship's view screen and proclaims that he is from the Earth of thousands of years in the past, and is eager to welcome "his children" to him. When Kirk asks and then demands that the *Enterprise* be released the face on screen threatens to "close his hand" and thus destroy the ship. Leading a landing party to the surface of the planet, with the mythologically significant name of Pollux IV, Captain Kirk and his crew confront the man responsible for their capture. He is dressed in white robes, is extremely muscular, is much larger than the crew of the *Enterprise*, and looks for all intents and purposes like an extra from a Greek or Roman Hollywood movie. And he claims at once that he is the Greek god Apollo. When Ensign Chekov replies, "And I am the tsar of all the Russias!" then apologizes to his captain for the outburst by saying, "I'm sorry, Captain. I never met a god before," Captain Kirk retorts, "[A] nd you haven't yet." The point is clear and one that is inherent in much of science fiction: when confronted by a superior force or a superior technology, look for the logical and empirical—not the supernatural—explanation. Or, as Kirk puts it to Apollo at one point, "you'll find we don't bow to every creature who happens to have a bag of tricks." And searching for the logical explanation is exactly what the crew of the *Enterprise* does for the remainder of the episode.

What follows is a story of Kirk and his crew working to escape both the force that holds their ship in space and the character who appears to control that force. Apollo repeatedly demands the worship of the human crew he has brought from the *Enterprise* and Kirk repeatedly denies him that worship. Scanning Apollo, Dr. McCoy claims he is fully humanoid, although he towers over the crew of the *Enterprise*. Eventually, using all the technology available to them through their ship, Kirk and his crew are able to fire their weapons and reduce the "god" to a "man-sized being." Like many of the episodes in the original series, the philosophical statement comes at the very end. On having his powers taken from him by the technology of the *Enterprise* Apollo, clearly anguished, proclaims, "I would have cherished you, cared for you. I would have loved you as a father his children! Did I ask so much of you?" Kirk replies with the message the episode intended, and it is a message intended by the franchise as a whole. Speaking not just to the god, Apollo but also presenting the franchise's view of supernatural or mythical explanations, Kirk declares, "We have outgrown you." And because the crew members of the *Enterprise* are more advanced, presumably, than the viewers of the television show he drives the point home, saying, "You asked for something we can no longer give."

The point of the episode is that the myths of the past have little to offer the star-faring crew of the *Enterprise* in the future—or the viewers of the show in the late 1960s. Kirk acknowledges, at the very end of the episode, "they gave us so much—the Greek civilization, much of our culture and philosophy came from the worship of those beings." But at the same time worship is no longer possible—or reasonable—in the world of *Star Trek*. John Shelton Lawrence has offered an analysis of this episode, noting this antimythic "conviction of Captain Kirk fits the spirit of the entire series. It is unthinkable that he or his crew, not to mention the strictly scientific Spock, would give credence to myths for a moment" (2010, 99).

If *Star Trek* has any moment of genius it is found in this very neat narrative trick. Where NASA was forced by the political rhetoric used to garner support of its logos-centered mission to make the lunar landings conform to certain mythological necessities, *Star Trek* offers almost the exact opposite format. By denying that there is a mythic content to the series, and by specifically undermining mythic narratives in many episodes, *Star Trek* can claim that the mythos it does present is, actually, an extrapolation of contemporary logos. Thus, where NASA attempted to make its logos act as mythos, *Star Trek* merely pretends its mythos is, in fact, logos. As a result the myths presented in the original series can be left largely unexamined and are taken as merely logical extensions of the current logos known to its viewers. Or, to put it simply, in the America of the 1960s and 1970s, pretending your mythos is logos as *Star Trek* did is a far easier mission than making your logos conform to a preexisting mythos as NASA attempted.

Myths are repudiated within many of the original episodes, especially myths of Eden or paradise. But it is done so wholly through the employment of another set of myths, the first being the myth that there simply are no longer any myths. Robert Jewett and John Shelton Lawrence have traced this unique presentation of *Star Trek*'s mythic qualities to this underlying "myth of mythlessness." Jewett and Lawrence see one of the prevailing myths of contemporary society as a myth that there are, indeed, no myths left. For Jewett and Lawrence the contemporary "claim to be antimythical and purely scientific is itself a myth" (1977, 17). *Star Trek*, in presenting itself as a nonmythic narrative, has thus allowed the unique position of being allowed to be mythic while at the same time being capable of claiming itself antimythical. Indeed, in describing the "myth of mythlessness" Jewett and Lawrence use *Star Trek* as their main example. For Jewett and Lawrence, then, within *Star Trek*'s narrative

> the mythical formulas so crucial to the plots are never called into question. Indeed, the myth of mythlessness ensures that they not even be acknowledged. Instead of a rigorously self-critical scientific outlook, *Star Trek* offers *pseudo-empiricism*, an empirical veneer of gadgetry, and crew talk applied to a mythical superstructure.
>
> (1977, 19; original emphasis)

But it is in refusing to acknowledge, or reflexively examine, those myths that *Star Trek* does use as the basis of its narrative the show offers a striking image of those myths that are then accepted by the viewer not mythically but as mere logical extensions of their contemporary lives. Yet, as Jewett and Lawrence have noted elsewhere, *Star Trek* is a narrative "extending from the older cowboy stories central to the [mythic American] archetype" (2002, 229). It is a franchise that works because its myths are well hidden in "pseudo-empiricism" and because it continually maintains a "fidelity" to the American mythic system of "plot, character, and communities in peril . . . [while] benignly joining science with redemptive tasks" (Lawrence and Jewett 2002, 224–225). To fully accomplish this *Star Trek*'s mythic structure must be almost exclusively American in nature, presenting an ongoing story of Starfleet officers as "superheroes who must bypass the restraints of law to redeem" all of the aliens they encounter (Jewett and Lawrence 2003, 35). This mythic trend will become significant in *Star Trek*'s examination of the Vietnam conflict and in its continuing rejection of any kind of paradise or Eden.

It is through this perspective that two issues, which bear analysis here, become most important. Other scholars, in varying degrees, have analyzed both though not in the fashion offered here. While during its initial three-year network run *Star Trek* dealt with both the Vietnam conflict and with the theme of paradise or Eden, but the show's narrative reactions to these two themes are strikingly divergent. Concerning Vietnam the narratives changed significantly throughout the series run, beginning with unqualified claims of support for imperialist interventions in Southeast Asia but ending with far more critical approaches. Where Kottak notes that, overall, *Star Trek* proposed as one of its "constant messages" that "strangers, even enemies, can become friends," when dealing with Vietnam it took the writers and producers of the series somewhat longer to vest that ideology into the current international relations of their home nation (1990, 104). Contrarily, when approaching the theme of paradise that NASA was also heavily invested in during the same period, *Star Trek*'s creative team offered a far more consistent approach. Time and time again Captain James T. Kirk rejects, as an outright necessity, any possibility of paradise.

It is with these two themes—paradise and Vietnam—that *Star Trek* offers its coherent version of the frontier myth. And, perhaps ironically, in doing so it offers a frontier myth that is far more *realistic* than either of those offered by the Vietnam conflict abroad or NASA at home in late 1960s America.

PARADISE RENOUNCED: EDEN AND PROGRESS IN *STAR TREK*

If NASA floundered in part because the rhetoric and myths used to propel their missions were destined to fail when no real Edenic "new world" was found in the depths of space, *Star Trek* offered a fully inverted form of that myth. Rather than imply that their mission was to seek such a paradise, the crew of the USS *Enterprise*, when they found such a world, would literally stop at nothing to change it. If much of American culture is, as Mircea Eliade explained, an ongoing dialogue about "the progressive transformation of the 'American Paradise,' giving rise to the myth of indefinite progress [and] to American optimism" then *Star Trek* fully dispenses with the need for the "paradise" to begin with in favor of, simply, "progress" (1969, 90). Captain Kirk is, to use Eliade's phrasing, "a modern man . . . irremediably identified with history and progress," but where Eliade sees this position as being identified with the concept that "history and progress are a fall, both implying the final abandonment of the paradise," Kirk and the crew of the *Enterprise* quite willfully reject paradise (1954, 162). This rejection of paradise is established, in part like all science fiction, by dealing with new technologies, encounters with alien "Others" to promote an expanded worldview, and the effects of these two narrative structures on the humans of the stories. Or, more specifically, progress and the future are more important than any paradisiacal past could ever be. To establish this, *Star Trek* spent at least ten of the original seventy-nine episodes confronting perfect, Edenic or utopian worlds and societies. This is far more episodes than feature the now-famous Klingon or Romulan aliens.[10] Karin Blair, in her classic essay provocatively titled "The Garden in the Machine: The Why of *Star Trek*," has argued that the Edenic episodes of the series intentionally reference "our cultural stereotype of paradise." Because of *Star Trek*'s method of dealing with such narratives, Blair argues—and intentionally references Leo Marx's phrase—"the television series moves beyond cultural stereotypes to archetypes and [offers] new possibilities for the future" (1979, 311–312). In almost every case what Blair calls "new possibilities for the future" is really defined by the possibility of social progress, and paradises, in that they deny such progress, by their very nature, are always shown to be actually dystopian. "It is no accident," Jon Wagner notes, "that people who live in Edenic 'paradises' [in *Star Trek*] are depicted as anesthetized and stagnated" (1999, 137). In almost every case the reaction of the *Enterprise* crew, or at least of Captain Kirk, to such stagnation is to find some way to both criticize such an Edenic world and to use that critique as a justification to alter it. Wagner and Jan Lundeen explain the logic of this necessity clearly:

> Utopias, according to these classic *Star Trek* narratives, are not only stagnant but tyrannical as well. Insofar as perfection requires orderly planning, it requires either voluntary submergence of the individual mind into that of the group and/or the rule of some oligarchy, dictator, machine, or other authority.
>
> (1998, 127)

In the "final frontier" of *Star Trek* progress is such a necessary good that, without it, members of any society are actually living in a false paradise—by whatever name it might be called. Because of the classic western tropes that are the basis of the series *Star Trek* "focused on the quintessential American dilemma of social order versus individual freedom" over and over again (Wagner and Lundeen 1998, 126). And, because this "dilemma" is something easily examined through an encounter with a perceived Edenic paradise, *Star Trek* turned to that theme, also, again and again. This narrative is so prevalent in the series that Wagner and Lundeen have provided a general narrative structure that applies to all of the original series' Edenic episodes:

> (1) A utopian—or rather, dystopian—community of peace and harmony is encountered on a newly discovered planet; (2) members of the *Enterprise* crew are drawn into the midst of this social order and (3) find their freedom jeopardized until (4) they finally manage to reemerge from this constricting social order by asserting their freedom of thought and action and (5) liberate the members/victims of the dystopia, (6) returning them to a more "natural" path of development, justified in (7) a concluding soliloquy about the shortcomings of "Paradise."
>
> (1998, 123)

Unlike NASA's rhetoric, which implied that a new paradise might await astronaut pioneers, *Trek*'s shift to the rejection of such a paradise is a major shift in American myth. Each of the three seasons of *Star Trek* contains multiple episodes following Wagner and Lundeen's pattern. However, an examination of just some of these episodes serves to highlight how that narrative is presented within *Star Trek*. In part this is because the theme is remarkably consistent throughout the three seasons of the show. But, more important, it also emphasizes the importance of those instances when the pattern is not followed.

In season one's "The Return of the Archons" the *Enterprise* is searching for the lost starship *Archon*, missing for over one hundred years. Exploring the planet Beta III crewmembers begin to act strangely, and Captain Kirk eventually leads a larger landing party to the planet to investigate. The natives of Beta III look entirely like humans, but their town and their clothes strongly resemble a Hollywood production placed in the late-nineteenth-century American Midwest. The natives exhibit little individuality and all behave in an overly polite fashion to the *Enterprise* crew people they encounter proclaiming things like "Peace be with you, brother." Quickly recognized by locals as not being "of the Body" Kirk must convince some of them to protect his landing party from "Landru," an apparently all-seeing dictator of some kind. In an effort to avoid capture, Kirk, who has surmised that this is a society based wholly on obedience, begins to refuse first requests and then orders from the agents of Landru. This sows enough confusion that, for the time being at least, Kirk and his party are able to escape capture.

One of the natives, Reger, does offer Kirk and company sanctuary, as well as an explanation for Landru. Landru, they are told, frequently absorbs people "into the Body" in order to facilitate proper behavior and right thinking. Eventually, Landru will utilize a kind of telepathic connection between all those "of the Body" to facilitate the capture of Kirk and his party. Chased through the town by the slowly moving, almost zombie-like natives (and some former *Enterprise* crewmen), the *Enterprise* party is finally incapacitated by some kind of ultrasonic wave and captured. In custody, but, of course, planning escape, Kirk is taken to the "Absorption Chamber," where he is to become part of "the Body" which is presented as a loss of his individuality and freedom of choice. But a secret resistance movement has infiltrated Landru's organization and helps Kirk to escape.

Kirk and Spock are able to deduce from what evidence they have exactly what has been happening on Beta III. Spock explains: "This is a soulless society, Captain. It has no spirit, no spark. All is indeed peace and tranquility. The peace of the factory, the tranquility of the machine. All parts working in unison." Spock's logic, as always, is impeccable and together with Kirk they quickly reach a conclusion: Landru is a computer imposing order and peace by eliminating freedom and progress. "Mr. Spock," Kirk declares, "the plug must be pulled!" After some work Kirk and Spock are able to confront the computer Landru in person, in a hidden chamber near the Absorption Chamber. Wagner and Lundeen's soliloquy is found partially in Kirk's confrontation with the computer Landru, and partially after the episode's climax. Landru explains to Kirk that its mission is to provide the natives with life, peace, harmony, and to protect them from evil. Through a logical game Kirk is able to convince Landru that its very existence is compromising its own stated goal. "Without freedom of choice, there is no creativity," Kirk tells Landru. "Without creativity, there is no life." As the computer begins to hesitate and smoke, Kirk finally delivers his coup de grâce, almost yelling, "You are the evil! The evil must be destroyed!" With that Landru self-destructs, and the static but harmonious life on Beta III is changed forever.

Kirk is able to tell Reger, as his landing party is returning to their ship, what their world will be like now. With Landru gone Reger is hopeful for freedom for himself and his people, and Kirk tells him, "It's time you learned that freedom is never a gift. It has to be earned." Back aboard the *Enterprise* Kirk and Spock are able to reflect on the events on Beta III. Reflexively Spock asks, "How often mankind has wished for a world as peaceful and secure as the one Landru provided?" And, building from the confrontation with the computer Landru, through his parting words to Reger, Kirk is able to finalize the lesson of the episode: "Yes. And we never got it," he says. "Just lucky, I guess."

The message is simple, and consistent through most of *Star Trek*. And each time it follows Wagner and Lundeen's narrative structure. Landru's efforts to provide peace, harmony, and safety lead inevitably to a loss of freedom, self-determination, and finally, progress. The natives of Beta III may have avoided conflict, but the cost was to lose change and, with it, progress as well. For the crew and the captain of the *Enterprise* this is clearly a devil's bargain leading not to Eden but to dystopia.

This message is made even more specific in another season one episode, "This Side of Paradise." The "natives" of the episode are actually Federation colonists on the planet Omicron Ceti III, but otherwise, Wagner and Lundeen's formula is followed exactly. The *Enterprise* and her crew do not expect to find anyone on the planet as, after the colony ship departed Earth it was discovered that the planet was regularly bombarded with deadly "berthold rays." But find the colonists they do, and all are alive, in perfect health, and delighted with their lives on the planet that should have killed them. It turns out the planet contains an alien flower that, after infecting humans with its spores, is capable of providing them with a sense of well-being while simultaneously protecting them from the berthold radiation.

As Kirk's crew quickly succumbs to the flower spores, the colony leader, Sandoval, explains the planet to Kirk. The captain is attempting to understand, first, why the colonists have survived and, second, why they are refusing evacuation from the harmful planet. Holding a handful of soil he proclaims, "This is the reason, Captain. This soil will grow anything we plant in it. It's a perfect world; it gives us all we need. It is perfect." As more and more of the *Enterprise* crew members become infected with the spores all begin to passively refuse orders and, eventually, mutiny against Kirk. Soon every crewmember of the *Enterprise* but Kirk will be infected, including the ever-logical Spock, who will even proclaim his love to one of the colonists while influenced by the spores, so powerful a force are they. Later, Spock will elaborate on the crew's mutiny and the perfection of their environment, saying to Kirk, "It's a true Eden, Jim. There's belonging and love." Kirk's reaction is fully in line with the philosophy of *Star Trek*. He replies to Spock, "No wants. No needs. We weren't meant for that. None of us. Man stagnates if he has no ambition, no desire to be more than he is."

Although eventually infected himself, Kirk will overcome the infection because of the great anger the thought of giving up his ship and living with the colonists on the planet below arouses in him. Powerful emotions are the key to counteracting the infection, and after inducing them in Spock as well, the two free the rest of the *Enterprise* crew and colonists. When freed from the infection of the spores colonist leader Sandoval is almost immediately disgusted with that has occurred in the years his people spent under their influence on the planet's surface. Looking around him at the undeveloped planet he exclaims, "We've done nothing here. No accomplishments, no progress. Three years wasted. We wanted to make this planet a garden." When Dr. McCoy then asks him what he and his fellow colonists want to do after they are evacuated elsewhere Sandoval replies with conviction, again providing the *Star Trek* philosophy: "I think I'd . . . I think we'd like to get some work done. The work we started out to do."

As if the message is not quite clear enough first with Kirk declaring a need for ambition to avoid stagnation, and Sandoval evincing a desire to leave the Edenic planet and "get some work done," the final exchange between Dr. McCoy and Kirk on the bridge of the *Enterprise* makes the message even more obvious. Reflecting on what has happened McCoy says, "Well, that's the second time man's been thrown out of paradise." Kirk's reply encapsulates an entire philosophy not only about paradise, or Utopia, or Eden, but about humanity as well:

> No, no, Bones. This time we walked out on our own. Maybe we weren't meant for paradise. Maybe we were meant to fight our way through. Struggle, claw our way up, scratch for every inch of the way. Maybe we can't stroll to the music of the lute. We must march to the sound of drums.

The message remains that paradise, however defined, leads to stagnation and that, in the universe of *Star Trek*, stagnation is to be avoided at all costs. It is far better to work toward bettering the universe than it is to exist, without real freedom, on an Edenic planet. Wagner and Lundeen do provide an eminently useful phrase in describing the way the final message of each of these episodes is presented when they call them Kirk's "concluding soliloquy about the shortcomings of 'Paradise.'" In the second-season episode "The Apple" the *Enterprise* crew visits the planet Gamma Trianguli VI, yet another paradisiacal planet under the control of a machine. Spock, Dr. McCoy and Kirk discuss the planet's wonderful temperature, excellent soil quality, and Kirk even declares, "It is spectacular, isn't it?" So perfectly suited to humans is the planet that when Ensign Chekov declares, in a small fit of homesickness, that it reminds him of his native Russia. McCoy can reply, "More like the Garden of Eden, Ensign." Like the many other such episodes, the natives live a life of ease with all of their needs provided for them by a being called "Vaal," which is depicted as a dragon-headed monster built into the side of a mountain. Practically immortal, the natives have no knowledge of death or sex and merely need to "feed Vaal" by depositing rocks into its mountainous gullet to provide energy for its computations. It is not likely a mere coincidence that Vaal's name bears a close resemblance to the Baal worshiped by the Israelites as they made their way out of bondage in Egypt on their way to Israel. For Vaal, Kirk will eventually prove, is equally a false god, and what it provides equally harmful.

In this episode it is McCoy who provides the summation of the argument against this paradise, presenting his argument to Captain Kirk:

> Jim, you're not just going to stand by and be blinded to what's going on here. These are humanoids, intelligent. They need to advance and grow. Don't you understand what my readings indicate? There's been no progress here in at least ten thousand years. This isn't life. It's stagnation!

Eventually, this argument is used to justify the destruction of Vaal and Kirk can introduce the planet's natives to life outside of Eden. This means, quite rightly, knowledge of life and death, as well as knowledge of sex and reproduction. In short, Kirk offers them what Wagner and Lundeen call "a path of development more palatable to Kirk's taste" (1998, 123). One of the natives complains about this change initiated by the captain, saying "But it was Vaal who put the fruit on the trees, caused the rain to fall. Vaal cared for us." Kirk replies with another soliloquy of *Star Trek*'s philosophy of progress:

> You'll learn to care for yourselves, with our help. And there's no trick to putting fruit on trees. You might enjoy it. You'll learn to build for yourselves, think for yourselves, work for yourselves, and what you create is yours. That's what we call freedom. You'll like it, a lot. And you'll learn something about men and women, the way they're supposed to be. Caring for each other, being happy with each other, being good to each other. That's what we call love. You'll like that, too, a lot. You and your children.

While the natives do not, as yet, even know what "children" are, the crew of the *Enterprise* can leave with the sure knowledge, as McCoy claims in the final scene of the episode, that "[w]e put those people back on a normal course of social evolution. I see nothing wrong in that."

The theme of Eden being more dangerous than it is paradisiacal is so central to original series episodes that, in the third season's "The Way to Eden," a roving band of what can only be called alien "space hippies" take control of the *Enterprise* to search for mythical planet that is literally named "Eden." The planet Eden is a hoped for return, as Karin Blair suggests, to "a cultural prototype of paradise" (1979, 313). The hippies expect a world with no work and ongoing ease of life. Like Kirk's encounter with the natives' claim in "This Side of Paradise" that "Vaal put the fruit on the trees," the space hippies sing about Eden as a place where they are "gonna live like a king on whatever I find/ eat all the fruit and throw away the rind." It will prove an impossible dream. While Kirk is allowed no closing soliloquy, the message is plain enough when, upon arriving on the planet Eden one of the space hippies, unsurprisingly named Adam, eats fruit of the Garden of Eden, which, it turns out, is poison. Indeed, all life on Eden is acidic and deadly to humanoids. Adam immediately dies because the sought-after Eden might be, again in Blair's phrase, a place "where there is no labor, no competition or need for order, no self-consciousness or responsibility." But Eden also "proves uninhabitable" (Blair 1979, 314).

And so, with Eden a place unfit for human life, the *Enterprise* can leave the planet for its next mission.

For Wagner and Lundeen the theme of paradise is far more central to *Star Trek* than any other issue the series ever tries to explore. The final verdict of *Star Trek*'s examination of paradise over three seasons and seventy-nine episodes is, "that the cost of such life is prohibitively high and that once in Paradise we will become trapped there (1998, 126) And, in each case Captain Kirk has no compunction about violating the Prime Directive, altering the societies he and his crew visit, and then vanishing into the distance for next week's episode.

This makes it all the more significant when Kirk encounters a "paradise" that he must alter and, within the narrative of the episode, is actually anguished and filled with self-doubt by the changes he initiates. More significant still, it is one of the few episodes in which *Star Trek* chooses to also grapple with an issue confronting all of their viewers: the Vietnam War. That the series should only offer a sense of loss when altering an alien Eden only when confronting the specific war that seemed to subvert American myths of such Edens is a key indicator that *Star Trek* was fully capable of interacting with the complexities of such myths.

STAR TREK AND THE WAR ON PARADISE

Though not speaking directly to *Star Trek*'s varied reactions to Vietnam, Ace G. Pilkington has noted that the show presented a "multiplicity of mythic messages and the many mirrors they have provided for America" (2010, 54). The possibility of so many "mythic messages" allowed the writers of *Star Trek* to offer what American ambassador and political scientist Mark P. Lagon called an "exploration" of "the moral dilemmas faced by American foreign policy makers" throughout the show's three-year run (1997, 234).

Lagon forwards the idea that Vietnam was one of those major "dilemmas" and that "the certainty with which Kirk entered into conflicts on planets and the ambivalence expressed by Spock about the logic of doing so" carry on the "debate among 'the best and the brightest' elites in the Kennedy and Johnson Administrations about United States involvement in Vietnam" (1997, 237). Because of *Star Trek*'s use of this "multiplicity of mirrors"—and the complexity of the "dilemma" of the war in South East Asia—the series' reaction to the Vietnam War was both variable and evolving. And it is because of the mythic dimensions of *Star Trek*, Lawrence and Jewett note, that it was "an effective format for reinstating in the realm of fantasy some of the American values that floundered in the 1960s against ugly obstacles in Vietnam" (2002, 213). At the same time, *Star Trek* was able to utilize those "mythic messages" for the purpose of telling different stories, with different moral lessons, about the American involvement in that war. This is both a strong indicator that *Star Trek* must be considered myth, and that myth remains a generally polysemous construct for telling moral and ethical lessons.

As a result, during the course of its three years in the 1960s *Star Trek* was able to move from generally positive representations of interventionist policies in Vietnam, to generally negative and then fully negative. A number of scholars have offered readings of those episodes in which *Star Trek* deals with the topic, and those readings have been largely concerned with the sociopolitical meanings inherent in the episodes under examination. Early in the show's run the message presented on Vietnam was very much about the unfortunate necessity of intervention, while later the show became far more idealistic in its negative message about intervention. Perhaps most important, however, is *Star Trek*'s presentation of Vietnam War themes in an episode that also offered the series' major narrative structure: a story about paradise. Examining first *Star Trek*'s interventionist apologetic, then its idealistic, turn makes this final episode all the more significant.

The first episode scholars turn to in discussions of *Star Trek* and Vietnam was aired at the very end of the first season of the series. Ostensibly about World War II, "The City on the Edge of Forever" is, in *Star Trek*'s usual method of allegorical storytelling, an episode about the awful necessity of war (Pevney 1967b). Written by well-known science fiction author Harlan Ellison, though changed through significant rewrites by various *Star Trek* producers, the final version of the script, according to H. Bruce Franklin, had "major plot element[s] whose subtext was the growing movement against the Vietnam War" (Franklin 1994, 26). Aired in early April of 1967, just as significant antiwar protests were beginning against the Vietnam War, the episode is a striking example of the early unreflexively positive statements made by *Star Trek* about the United States' contemporaneous political realities. A reverse time-travel episode, the plot revolves around Captain Kirk, Mr. Spock, and Dr. McCoy, all of whom find themselves in 1930s New York City and confronted with issues of war, peace, and intervention. The episode begins with Dr. McCoy accidentally injecting himself with a drug overdose, and delusional, he escapes the ship to the surface of a planet exhibiting "time distortions." There he uses a deus ex machina time-travel device called "The Guardian" to send himself to the past. Immediately on McCoy's jump into the device, the present for Kirk and Spock, now on the planet's surface as well, is altered. Indeed, their ship and the United Federation of Planets have ceased to exist. Knowing the standard time-travel trope of science fiction—that any change to the events of the past can alter the reality of the present—Kirk and Spock follow McCoy through the Guardian in an effort to both "fix" history and return the deranged doctor to their own present.

While searching for McCoy in 1930s New York, however, Kirk meets a beautiful social worker, Edith Keeler, whom he immediately falls in love with. As the narrative develops Kirk's love interest is revealed to be a pacifist and, because of the presence of three members of the *Enterprise* crew, will survive when previously she should have died. The result will be an antiwar movement in the United States lead by Keeler and delaying the nation's entry into World War II, thus allowing the Nazis to achieve victory and eventual world domination. And, as a result, the future—Kirk's and Spock's present—will never exist. Mr. Spock delivers the logical solution that will restore the "time line" and allow the USS *Enterprise* to exist: "Jim, Edith Keeler must die."

The episode ends with exactly that. Edith Keeler is killed by a truck as Kirk, Spock, and McCoy look on, the latter recovered from his overdose. Kirk is anguished by Keeler's death, and the ever-emotional McCoy, not yet aware of the significance, asks, "I could have saved her! Do you know what you just did?" Kirk is unable to reply, but Spock does, offering, "He knows, Doctor. He knows." For John Shelton Lawrence, much like Franklin, the message of the episode is clear. It is an "explicitly worked out" message in favor of both interventionist policies and against pacifism or "an apologetic for the Vietnam War" (Lawrence 2010, 97).

Coming in April 1967 places "The City on the Edge of Forever" right at the cusp of an awakening of antiwar feeling in the United States. H. Bruce Franklin has investigated both the production history of the episode, and discussed it with those involved with the show. His conclusion, that Vietnam was a major subtext of the episode, is confirmed by one of the original series' producers, Robert Justman. Franklin writes,

> In the episode as it aired in 1967, this speculation introduced into the June 3, 1966 script has been turned into a major plot element whose subtext was the growing movement against the Vietnam War. Asked in 1992 whether the makers of this episode consciously intended it to have the contemporaneous anti-Vietnam-war movement as subtext, producer Robert Justman replied, "Of course we did."
>
> (1994, 28)

For the writers and producers of *Star Trek*, many of whom had served in World War II, the message of the episode could not be clearer. Imagine, they asked, an antiwar movement in late 1930s New York. Not only would the world of their viewers be forever different, but the fictional world of *Star Trek* would have been lost as well. Fortunately, the "multiplicity" of these "mythic mirrors" would quickly move beyond such profoundly positive views of such interventionist policies.

"The City on the Edge of Forever" is, at best, an ambiguous statement about Vietnam. But where that episode works because it is such a general allegory most critics are in agreement about the more specific and vehemently anti–Vietnam War message of the second-season episode, "The Omega Glory" (McEveety 1968). Written by Gene Roddenberry and first aired in early March 1968 it is among the most didactic of original series episodes. First conceived as one of the possible pilot episodes for the series, it would not be produced and aired until considerable rewrites were done later, making it, as Nicholas Evans Sarantakes put it, "a patriotic but thoughtful piece of propaganda" (2005, 86). After finding the seemingly abandoned USS *Exeter* in orbit of the unexplored planet Omega IV, Kirk, Spock, McCoy, and an always-expendable extra lieutenant beam to the planet's surface. What they find on the planet is seemingly endless war between the Asiatic-appearing "Kohms" and the blonde, very occidental "Yangs." Initially the crew members find that they have more in common with the "Kohms," who seem rational, intelligent, and, when compared with the seemingly savage "Yangs," civilized. It soon becomes apparent that this is a mistaken assumption. As Daniel Leonard Bernardi describes it in examining the racist underpinnings of the episode,

> [a]s the story progresses, the Yangs are constructed as noble savages; their cause to annihilate the [Kohms] is established as justified. [The Kohms], on the other hand, are constructed as brutal and oppressive; their drive to suppress the Yangs is established as totalitarian. (1998, 58)

Kirk and Spock quickly deduce precisely who the two groups are when they realize that "Kohms" are "communists" and the "Yangs" "Yankees." All that remains to be uncovered is the "Omega Glory" and the "worship words" of the Yangs. The "Omega Glory" is, not at all surprisingly by the end of the episode, a tattered American flag and the "worship words" merely a jumble of the preamble to the U.S. Constitution and the Pledge of Allegiance that the Yangs no longer understand. From here it is merely the moral lesson that Kirk must deliver, and in the end he does precisely that. Standing before the Yangs he reads them their worship words—phrases that are meant to be spoken, the Yangs believe, only by the elite—while holding a native copy of the Constitution. But he goes further and offers not merely a recitation but outright ideology:

> Among my people, we carry many such words as this from many lands, many worlds. Many are equally good and are as well respected,

> but wherever we have gone, no words have said this thing of importance . . . in quite this way. Look at these three words written larger than the rest with a special pride never written before or since—tall words proudly saying "We the people." That which you called *Ee'd Plebnista* [The Pledge of Allegiance] was not written for chiefs or kings or warriors or the rich and powerful, but for all the people!

Kirk continues to recite much of the preamble of the Constitution to the Yangs, as they look at him with confusion. Finally, he declares that "these words and the words that follow were not written only for the Yangs, but for the Kohms as well! They must apply to everyone or they mean nothing!" Eventually Kirk convinces the Yangs of the seemingly inherent truth of both their worship words and of their universal application. H. Bruce Franklin has suggested that in "The Omega Glory," the seemingly propagandistic idealism of Kirk's final speech "displayed a profound darkening of *Star Trek*'s vision of the Vietnam War and its possible consequences" (1994, 29). For Franklin *Star Trek* was suggesting that in fighting an endless war against a foe for reasons that the Yangs did not understand or remember they were undermining their own ideals in precisely the same way as the United States was in Vietnam. He goes further:

> "The Omega Glory" implies that the war in Southeast Asia, which no longer held any promise of victory or even suggestion of an end, could evolve into an interminable, mutually destructive conflict between the "Yankees" and the "Communists" capable of destroying civilization and humanity. True Americanism is shown as antithetical to mindless militarism and anti-Communism, and the episode rather paradoxically uses ultrapatriotic images of a tattered Old Glory and strains of the Star Spangled Banner to preach a message of globalism
>
> (Franklin 1994, 32).

With "The Omega Glory" *Star Trek* would move to a far more idealistic and antiwar position concerning Vietnam. Although in the single episode in which the characters specifically reference Vietnam the message is far more ambiguous. Captain Kirk's behavior and justification for that behavior will seem, initially, supportive of the American involvement in Southeast Asia. However, the episode also presents yet another paradise that Kirk alters. However, this time Kirk is actually angered by the cultural intervention and worried about the ethics of such an intervention. That ethical concerns are featured in an episode that references Vietnam when normally the series has no such worries makes it all the more provocative a statement about the moral and mythic cultural effects of that conflict.

The episode is "A Private Little War," and in it the crew of the *Enterprise* investigates the planet Neural, which Captain Kirk had visited thirteen years earlier as a lieutenant aboard the Starship *Farragut* (Daniels 1968). An old friend of Gene Roddenberry, Don Ingalls, originally wrote the episode as a profoundly negative attack on the American involvement in Vietnam. The episode would be rewritten by Roddenberry to such an extent that Ingalls would use the pseudonym Jud Crusis in the story credit as a protest (Sarantakes 2005, 90–91). Roddenberry's rewritten version of the episode would thus be aired in early February 1968, during the height of the Tet Offensive in Vietnam, strikingly at a turning point in both the war and the American population's opinion on that war.

"A Private Little War" does not attempt to hide the perspective it is offering on current American political and military involvement in Vietnam. On the contrary, it is an episode so driven by the allegory that it intends that it is nearly impossible to watch it without recognizing the intentional references to the Cold War and the American involvement in Vietnam. The episode begins on the planet Neural's pristine and tropical surface where Kirk, Spock, and McCoy are examining the vegetation with their instruments. Kirk describes the tracks of a local predator as being several days' old and not of concern. Establishing the precise narrative that will follow, Spock replies, "Aside from that, you say it's a Garden of Eden?" Kirk replies that it had seemed that way when he visited the planet "as a brash young lieutenant" and adds, "Except these people stayed in their Garden of Eden. Bows and arrows for hunting, but absolutely no fighting among themselves. Remarkably peaceful and tranquil."

The images of the planet suggesting social stability and an Edenic environment firmly establish the type of *Star Trek* episode this should be: it is a situation that would immediately be cause for changing either the environment or the social system. Soon, however, such concerns are no longer a problem. Kirk and his landing party find natives who, contrary to Kirk's earlier description, carry flintlock rifles—weapons that Kirk declares the natives could not have developed in the thirteen years since his last visit. What follows is an episode in which the crew of the *Enterprise* discovers that the main enemy of the Federation, the Klingon Empire, is supplying the weapons to one group of natives, "the villagers," in an effort to begin the process of taking the planet into their sphere of influence. H. Bruce Franklin has noted that "the villagers" "represent the official U.S. view of the North Vietnamese," while the contrasting "hill people" are clearly meant as a similar U.S. view of the South Vietnamese (1994, 28). Nicholas Evan Sarantakes takes such a reading a step further, declaring that both Ingalls's initial story outline and the eventual episode have a similar scope in that for in both the "tropical planet and Vietnam were important solely as symbols of commitment" in a Cold War narrative of geopolitical containment (2005, 91). However, when finally broadcast the story "had been reworked to suggest that the United States was attempting to do the right thing in a situation in which there really was no good course of action" (Sarantakes 2005, 94).

Coming up with something for that "no good course of action" would fall to Captain Kirk. Forced into a situation of interstellar "geopolitics" Kirk must decide what to do in order to contain the Klingon Imperial advancement. His solution, very much the early justification of the United States in Vietnam, is to supply the other side of natives with similar weapons. Kirk, talking with Dr. McCoy, makes the allegory as specific as possible. He asks, "Bones, do you remember the twentieth-century brush wars on the Asian continent? Two giant powers involved, much like the Klingons and ourselves. Neither side felt that they could pull out?" McCoy replies as if he is declaring the contemporary American public's cry about Vietnam: "Yes, I remember—it went on bloody year after bloody year!" Captain Kirk can then offer both his solution and what seems to be the episode's justification for American involvement in Southeast Asia. He says,

> But what would you have suggested? That one side arm its friends with an overpowering weapon? Mankind would never have lived to travel space if they had. No—the only solution is what happened, back then: a balance of power.

As a result of Kirk's actions, the Klingon advancement will be thwarted, and a balance of power maintained between the Klingon Empire and the Federation. The implication is that a similar strategy is at work for the United States in Vietnam. For Sarantakes this solution makes "A Private Little War" emblematic not just of the U.S. involvement in Vietnam but also of the broader cultural reasoning of the Cold War itself. H. Bruce Franklin is even more specific in drawing out the episode's message, declaring "A Private Little War" to be about apologizing for the United States' "imperial interests in Southeast Asia":

> This is not to say that the episode implicitly endorses major enlargement of the Vietnam War. Indeed, it seems to suggest that the main danger to be avoided is any form of military intervention that could lead to direct warfare between the United States, here represented by the Federation, and the evil Communist empire, here of course represented by the Klingons.
>
> (1994, 31)

American film and television scholar Rick Worland is somewhat more forgiving, however, calling the allegory of Vietnam "ambiguous," and notes that when confronted with "no good course of action" Captain Kirk, in this episode at least, "seems, if not exactly critical, greatly anguished by [his] role in the loss of this Eden to bloodshed and treachery and about the implications of what is done to maintain the balance of power" (1988, 114).

Kirk's anguish is not subtle. At the end of the episode, as he is preparing to supply the native "hill people" with "one hundred flintlocks," engineer Scotty asks, "I didn't get that exactly, Captain. One hundred what?" Kirk, who for much of the run of the original series will have absolutely no qualms altering or undermining any Eden or paradise he and his crew encounter, replies, "A hundred serpents. Serpents for the Garden of Eden."

The Edenic nature of Neural is a key component to "A Private Little War." In every other episode in which Captain Kirk and the *Enterprise* crew encounter such an environment and society, they use their own internal logic to, first, thwart the Prime Directive and justify intervention and then, second, congratulate themselves for their action. In "The Return of the Archons" Kirk concludes that humans were "just lucky" to have never experienced paradise while, in "This Side of Paradise" that conclusion is taken a step further when he remarks, "maybe we weren't meant for paradise." In "A Private Little War," however, Kirk is fully aware that he is offering "serpents" to the natives, disrupting their way of life, all to preserve a "balance of power." This is not a Kirk who seems tired of his United Federation of Planets radically altering any paradise he encounters for the good of the natives. Rather, it is as if the producers, writers, and actors of an American television show seem exhausted by a war in Southeast Asia they have no control over. The final line of the episode is given to Kirk, and it might as well have been delivered by an actual American grunt in the rice paddies of Vietnam. With the unfortunate decision made, the balance of power seemingly established, and the Edenic society of Neural obviously to be destroyed, Kirk speaks to his communicator and says to his ship safely in space and away from the conflict he has just created: "We're very tired, Mister Spock. Beam us up home."

The movement in these three Vietnam episodes from an unreflexive support of American interventionism in "The City on the Edge of Forever," to what H. Bruce Franklin calls a "profound darkening of *Star Trek*'s vision of the Vietnam War and its possible consequences" in "The Omega Glory" and, finally, the exhaustion of Captain Kirk as he destroys yet another paradise in "A Private Little War" is not altogether surprising (1994, 32). The myths *Star Trek* employs, as William G. Doty has noted about all myths, are meant to "recognize [the] power to *seed creative new interpretations*," which is precisely what happens through these three episodes (2000, 418; original emphasis). If *Star Trek* writer David Gerrold is correct that science fiction, generally, is about "testing our culture," meaning "both the characters and the culture have to be recognizable parts of ourselves," then *Star Trek*'s three "Vietnam episodes" offer differing tests and different conclusions as well.

If the alterations in the outlook of these "tests" through these episodes are not surprising, however, what remains surprising is the treatment of the paradise described in "A Private Little War." In no other *Star Trek* episode will Captain Kirk indicate any qualms about altering or abolishing such a paradise. Indeed, if there is a single mythological trope that *Star Trek* faithfully adheres to throughout its original three years on television it is, simply, that Eden, or utopia, or paradise, is to be avoided, rejected, or destroyed, no matter what the consequences for those enjoying the paradisiacal conditions. Unlike the Kirk of "A Private Little War," the Captain Kirk who encounters other Edenic planets will not be anguished by his actions in abolishing them. If anything he will go to great lengths to justify his actions each time. Clearly, by the time of "A Private Little War," the producers of *Star Trek* are, at least unconsciously, aware of the mythic damage being caused by Vietnam. They can offer no solution, but they are able to express not just the political and military fatigue of the American people, but the mythic exhaustion as well.

Gene Roddenberry once noted that he had grand designs for *Star Trek*, but realistically, he offered, "assuming he [a television producer] conceived a program of such meaning and importance that it could ultimately change the face of America, he probably could not get in on the air or keep it there!" (Whitfield and Roddenberry 1968, 21). And, of course, he was correct. For all of Kirk's anguish and exhaustion at the destruction of yet another paradise cannot change one significant factor. While Roddenberry's attempt at creating a show of "meaning and importance" did, indeed, alter the American cultural landscape at best it only contributed to such change and never initiated it. *Star Trek*, like NASA, could not offer Marx's "new symbols of possibility" but merely a reworking of old symbols. And, finally, Roddenberry was not able to keep it on the air. Its final episode aired on June 3, 1969, just three weeks before Armstrong and Aldrin walked on the moon. And while the two astronauts were unable to provide their earthbound audience with the potential paradise the rhetoric that had supported the mission called for, the rejection of such paradises in *Star Trek* seemed to have ended as well.

At the same time while the crews of the Apollo missions were real, *Star Trek*, Captain Kirk, and the entire crew of the USS *Enterprise* remained wholly fictional. Where NASA's failure might be traced to rhetoric that did not originate with the agency, it is not a failure of *Star Trek*'s to note that its rhetoric existed primarily just to tell stories—to offer Doty's "forms of the future" in digestible, weekly installments.

If there is a bridge between the logos-driven world of the lunar missions and the mythos-powered hopes of *Star Trek*, neither the astronauts nor Captain Kirk would be able to cross it. For that mission what was needed was someone familiar with both sides of the equation. And that person would be failed astronaut, *Star Trek* fan, and Princeton University professor of physics Gerard K. O'Neill. What he offered was both the boundless opportunities of *Star Trek* and the implied paradises of NASA.

And what he called this combination of rhetoric, myth, and technology was "The High Frontier."

NOTES

1 The full litany, spoken by Kirk or others, can be found in most of the iterations of *Star Trek*. The original is worded as "Space: the final frontier. These are the voyages of the starship *Enterprise*. Its five-year mission: to explore strange new

worlds, to seek out new life and new civilizations, to boldly go where no man has gone before."

2 The "official" biography of Captain Kirk, unsurprisingly, is long and varied. With all character descriptions here, an attempt is made to stay within the confines of the narrative of the original 1966–1969 series. For details on both that version of Kirk and later versions see McVeigh (2010).

3 Boyd is discussing the later *Star Trek: The Next Generation*, but her summary of the Prime Directive is equally applicable to the original series.

4 Parts of this chapter introduction, with further details, can be found in Kapell (2010a, 214–215).

5 Originally published in the *Journal of Popular Culture* in 1977, Tyrrell is more readily available here in Kapell (2010d) as Tyrrell (2010b). For all works reprinted in Kapell (2010d) the citations will reflect the more recently published version.

6 Bernardi also offers a very useful bibliography of works discussing racism in *Star Trek*.

7 Similar to Bernardi on racism, Helford provides an excellent list of critical works on gender and sexism in the original series.

8 Malmgren is writing primarily of written science fiction, but his thesis is also applicable to film and television.

9 Geraghty (2008) offers an examination of the various drafts of the original opening narration that, while it does not specify the emergent imperialist paradigm, does imply it nicely. Eventually "no man has gone before" will be changed to "no one" in an attempt at gender neutrality.

10 Blair (1979) offers a list of such episodes that is useful here as many are discussed, in somewhat more detail, in this chapter and notes that other episodes could be includes with only slight changes in the definitions of *paradise*, *Eden*, or *utopia*. Blair's list includes Daniels (1966, 1967, 1968), Sparr (1966), Pevney (1967a, 1967b, 1967c); Senensky (1967a, 1967b), Taylor (1968, 1969), Leader (1968); and Alexander (1969).

WORKS CITED

Abbott, Carl. 2006. *Frontiers Past and Future: Science Fiction and the American West*. Lawrence: University of Kansas Press.

Alexander, David. 1969. "The Way to Eden." DVD. *Star Trek*. Los Angeles: NBC.

Bernardi, Daniel. 1998. *Star Trek and History: Race-ing Toward a White Future*. New Brunswick, NJ: Rutgers University Press.

Blair, Karin. 1979. "The Garden in the Machine: The Why of Star Trek." *The Journal of Popular Culture* 13 (2): 310–20.

Boyd, Katrina G. 1996. "Cyborgs in Utopia: The Problem of Radical Difference in Star Trek: The Next Generation." In *Enterprise Zones: Critical Positions on Star Trek*, edited by Taylor Harrison, Sarah Projansky, Kent A. Ono, and Elyce Rae Helford, 95–113. Boulder: Westview Press.

Daniels, Marc. 1966. "The Menagerie (Parts I and II)." DVD. *Star Trek*. Los Angeles: NBC.

———. 1967. "Who Mourns for Adonais?" DVD. *Star Trek*. Los Angeles: NBC.

———. 1968. "A Private Little War." DVD. *Star Trek*. Los Angeles: NBC.

Doty, William G. 2000. *Mythography: The Study of Myths and Rituals*. 2nd ed. Tuscaloosa: University of Alabama Press.

During, Simon. 2005. *Cultural Studes: A Critical Introduction.* New York: Routledge.

Eliade, Mircea. 1954. *The Myth of the Eternal Return.* New York: Pantheon Books.

———. 1969. *The Quest: History and Meaning in Religion.* Chicago: University of Chicago Press.

Franklin, H. Bruce. 1994. "Star Trek in the Vietnam Era." *Science Fiction Studies* 24 (1): 24–34.

Fulton, Valarie. 1994. "Another Frontier: Voyaging West with Mark Twain and Star Trek's Imperialist Subject." *Postmodern Culture* 4 (3): 1–24.

Geraghty, Lincoln. 2008. "Eight Days That Changed American Television: Kirk's Opening Narration." In *The Influence of Star Trek on Television, Film and Culture*, edited by Lincoln Geraghty, 11–21. Jefferson, NC. and London: McFarland & Co.

Gerrold, David. 1973. *The World of Star Trek (First Edition).* New York: Ballantine Books.

Goulding, Jay. 1985. *Empire, Aliens and Conquest.* Toronto: Sisyphus Press.

Helford, Elyce Rae. 1996. "'A Part of Myself No Man Should Ever See': Reading Captain Kirk's Multiple Masculinities." In *Enterprise Zones: Critical Positions on Star Trek*, edited by Taylor Harrison, Sarah Projansky, Kent A. Ono, and Elyce Rae Helford, 10–31. Boulder: Westview Press.

Isaacs, Bruce. 2010. "A Vision of a Time and Place: Spiritual Humanism and the Utopian Impulse." In *Star Trek as Myth: Essays on Symbol and Archetype at the Final Frontier*, edited by Matthew Wilhelm Kapell, 182–96. Jefferson, NC and London: McFarland & Co.

Jewett, Robert, and John Shelton Lawrence. 1977. *The American Monomyth.* Garden City: Anchor Press.

———. 2003. *Captain America and the Crusade Against Evil: The Dilemma of Zealous Nationalism.* Cambridge, UK and Grand Rapids, MI: WB. Eerdmans.

Kapell, Matthew Wilhelm. 2010a. "Conclusion: The Hero with a Thousand Red Shirts." In *Star Trek as Myth: Essays on Symbol and Archetype at the Final Frontier*, edited by Matthew Wilhelm Kapell, 213–19. Jefferson, NC and London: McFarland & Co.

———. 2010b. "Introduction: The Significance of the Star Trek Mythos." In *Star Trek as Myth: Essays on Symbol and Archetype at the Final Frontier*, edited by Matthew Wilhelm Kapell, 1–16. Jefferson, NC and London: McFarland & Co.

———. 2010c. "Speakers for the Dead: Star Trek, The Holocaust, and the Representation of Atrocity." In *Star Trek as Myth: Essays on Symbol and Archetype at the Final Frontier*, edited by Matthew Wilhelm Kapell, 67–79. Jefferson, NC and London: McFarland & Co.

———., ed. 2010d. *Star Trek as Myth: Essays on Symbol and Archetype at the Final Frontier.* Jefferson, NC and London: McFarland & Co.

Kottak, Conrad Phillip. 1990. *Prime-Time Society: An Anthropological Analysis of Television and Culture.* The Wadsworth Modern Anthropology Library. Belmont, CA: Wadsworth Pub. Co.

Lagon, Mark P. 1997. "'We Owe It to Them to Interfere': Star Trek and U.S. Statecraft in the 1960s and the 1990s." In *Political Science Fiction*, edited by Donald M. Hassler and Clyde Wilcox, 234–50. Columbia: University of South Carolina Press.

Lawrence, John Shelton. 2010. "Star Trek as American Monomyth." In *Star Trek as Myth: Essays on Symbol and Archetype at the Final Frontier*, edited by Matthew Wilhelm Kapell, 93–111. Jefferson, NC and London: McFarland & Co.

Lawrence, John Shelton, and Robert Jewett. 2002. *The Myth of the American Superhero*. Grand Rapids, MI: WB. Eerdmans.

Leader, Tony. 1968. "For the World Is Hollow and I Have Touched the Sky." DVD. *Star Trek*. Los Angeles: NBC.

Lichtenberg, Jacqueline, Sondra Marshak, and Joan Winston. 1975. *Star Trek Lives!* London: Corgi.

Malmgren, Carl. 1991. *Worlds Apart: Narratology of Science Fiction*. Bloomington: Indiana University Press.

McEveety, Vincent. 1968. "The Omega Glory." DVD. *Star Trek*. Los Angeles: NBC.

McVeigh, Stephen. 2010. "The Kirk Doctrine: The Care and Repair of Archetypal Heroic Leadership in J.J. Abrams' Star Trek." In *Star Trek as Myth: Essays on Symbol and Archetype at the Final Frontier*, edited by Matthew Wilhelm Kapell, 197–212. Jefferson, NC and London: McFarland & Co.

Mogen, David. 1993. *Wilderness Visions: The Western Theme in Science Fiction Literature*. 2nd ed., revised and expanded. Rockville, MD: Wildside Press.

Pauling, James Kirke. 1832. *Westward Ho! A Tale*. Vol. 1. 2 vols. New York: J & J Harper.

Pevney, Joseph. 1967a. "The Return of the Archons." DVD. *Star Trek*. Los Angeles: NBC.

———. 1967b. "The City on the Edge of Forever." DVD. *Star Trek*. Los Angeles: NBC.

———. 1967c. "The Apple." DVD. *Star Trek*. Los Angeles: NBC.

Pfitzer, Gregory M. 1995. "The Only Good Alien Is a Dead Alien: Science Fiction and the Metaphysics of Indian-Hating on the High Frontier." *Journal of American Culture* 18 (1): 51–66.

Pilkington, Ace G. 2010. "Star Trek: American Dream, Myth, and Reality." In *Star Trek as Myth: Essays on Symbol and Archetype at the Final Frontier*, edited by Matthew Wilhelm Kapell, 54–66. Jefferson, NC and London: McFarland & Co.

Richards, Thomas. 1997. *Star Trek in Myth and Legend*. London: Orion Media.

Sarantakes, Nicholas Evan. 2005. "Cold War Pop Culture and the Image of U.S. Foreign Policy." *Journal of Cold War Studies* 7 (4): 74–103.

Senensky, Ralph. 1967a. "This Side of Paradise." DVD. *Star Trek*. Los Angeles: NBC.

———. 1967b. "Metamorphosis." DVD. *Star Trek*. Los Angeles: NBC.

Sparr, Robert. 1966. "Shore Leave." DVD. *Star Trek*. Los Angeles: NBC.

Star Trek Producers. 1967. "The Star Trek Guide (Writer's Bible)."

Steckmesser, Kent Ladd. 1997. *The Western Hero in History and Legend*. Norman: University of Oklahoma Press.

Taylor, Jud. 1968. "The Paradise Syndrome." DVD. *Star Trek*. Los Angeles: NBC.

———. 1969. "The Mark of Gideon." DVD. *Star Trek*. Los Angeles: NBC.

Tyrrell, Wm. Blake. 2010a. "Afterword: 409 Words on 'Star Trek as Myth and Television as Mythmaker.'" In *Star Trek as Myth: Essays on Symbol and Archetype at the Final Frontier*, edited by Matthew Wilhelm Kapell, 27–28. Jefferson, NC and London: McFarland & Co.

———. 2010b. "Star Trek as Myth and Television as Mythmaker." In *Star Trek as Myth: Essays on Symbol and Archetype at the Final Frontier*, edited by Matthew Wilhelm Kapell, 19–28. Jefferson, NC and London: McFarland & Co.

Wagner, Jon. 1999. "Intimations of Immortality: Death/Life Mediations in Star Trek." In *Star Trek and Sacred Ground: Explorations of Star Trek, Religion, and American Culture*, edited by Jennifer E. Porter and Darcee L. McLaren, 119–38. Albany: State University of New York Press.

Wagner, Jon, and Jan Lundeen. 1998. *Deep Space and Sacred Time: Star Trek in the American Mythos*. Westport, CT: Praeger.

Ward, John William. 1955. *Andrew Jackson: Symbol for an Age*. New York: Oxford University Press.

Westfahl, Gary. 1996. "Where No Market Has Gone before: 'The Science-Fiction Industry' and the 'Star Trek Industry.'" *Extrapolation* 37 (4): 291–301.

Whitfield, Stephen E., and Gene Roddenberry. 1968. *The Making of Star Trek*. New York: Ballantine Books.

Worland, Rick. 1988. "Captain Kirk: Cold Warrior." *Journal of Popular Film and Television* 16 (3): 109–17.

6 The High Frontier of Gerard K. O'Neill

An Endless Frontier Utopia in Orbit

America is a frontier society lacking a frontier.

—Gerard K. O'Neill, 1981 (1981, 71)

Princeton University physicist Gerard K. O'Neill wanted to be an astronaut.

By the summer of 1969, as *Star Trek* aired its last original episode, more than half a million Americans served in the Vietnam theater of operations, and Armstrong and Aldrin prepared to land on the moon, another version of the frontier was being planned. Gerard K. O'Neill's version of the frontier would bring together both traditional and new ideas of frontier mythology in the teaching and professional life of a failed astronaut and professor of Physics in both the summer and fall of 1969—and then throughout the 1970s. That fall his Physics 103 students would be asked to consider a new frontier in orbit, which O'Neill called the High Frontier, as a solution to all the maladies sffecting humankind. Be it pollution or population, energy or equality, the High Frontier was *the* solution. After the fall 1969 semester O'Neill reworked his ideas about the High Frontier in multiple publications throughout the 1970s merging technology, space travel, and frontier colonization into a grand utopian synthesis whose popularity serves as a strong indicator of the importance of frontier rhetoric in American myth and culture.

The NASA version of the American frontier myth would play a significant role in Gerard K. O'Neill's attempt to argue for orbiting space colonies as a new frontier. But he would also draw on a utopian tradition of the frontier as a place of endless possibilities that resonated with *Star Trek* while at the same time picturing that frontier as something akin to suburbia in space. O'Neill's frontier exceeded the myths of nineteenth-century American frontier and the reworking of that frontier by either NASA or *Star Trek*. As the American cultural studies scholar De Witt Douglass Kilgore notes, in O'Neill's High Frontier, "building orbital space colonies [would] allow the human race to avoid the dangers of overpopulation, pollution, famine, war, and tyranny" (2003, 159). What O'Neill saw in his orbiting space colonies was nothing less than a truly endless frontier. Speaking to the World Future Society meeting in Washington, D.C., in 1975, he put it bluntly: "The human race stands now on the threshold of a new frontier, whose richness surpasses a thousand fold that of the new Western world of five hundred years ago" (O'Neill 1977, 8).

Beginning at Princeton University in the fall semester of 1969 and expanded throughout the 1970s O'Neill, in a series of articles and books, put forth an image of millions of people living in space. Located in gravitationally stable orbits in the Earth–Moon system, and built with raw materials mined from the surface of the Moon, the colonies would offer low-cost solar energy to the Earth while simultaneously offering the traditional "safety valve" for the growing population on the home planet. Built near the colonies would be orbiting factories, allowing for a movement of pollution-producing industry from the surface of Earth as well. Added to this would be, in the classic Turnerian scheme of the mythic American frontier experience, the development of new and experimental social systems.

In short, O'Neill claimed his High Frontier would be a frontier utopia.

Eric Drexler, then a twenty-two-year-old PhD student in physics at the Massachusetts Institute of Technology who had, at the time, been active in promoting the idea of space colonies for seven years, outlined a history of O'Neill's idea in 1977. First offering credit and thanks to "sociologists, anthropologists, engineers, biologists, artists, and celestial mechanicians" for their contributions to the concept, Drexler then presented an overview of O'Neill's 1977 version of the High Frontier concept:

> What has emerged after yet another year of refinement, criticism, and invention is a gradual program involving 100–200 flights of the space shuttle to set up a system capable of getting rock off the Moon and refining it into metals. After that, [solar] power satellites would be built to supply energy to the Earth, and small colonies would be built to house the people building power satellites.[1]

The "small colonies" Drexler refers to consisted of hypothetical populations in the early 1990s in the tens of thousands and not too much later than that in the millions. It might appear a hyperbolic statement that Drexler could include in his byline for his history of the High Frontier, "I probably won't die on this planet," but in 1977 the idea had already gained enough momentum that for a NASA in search of a post-Apollo mission it was worth a major design study (Drexler 1977, 104; Johnson and Holbrow 1977). That study concluded, optimistically, that, after a "thorough and comprehensive . . . 10-week summer study" project on the subject, "the obstacles to further expansion of human frontiers in this way are principally philosophical, political, and social rather than technological" (Johnson and Holbrow 1977, 1). Materials for building would, first, be mined on the Moon and, later, in the asteroids between Mars and Jupiter. Launched to build orbital cities via a "mass-driver" system of magnets, the materials would be used to first build an "Island One" and then solar power satellites beaming energy to Earth to help pay for continued operations and building of larger habitats in orbit (O'Neill 1976b, 139–146).[2] The cheap solar power would then be used to move industrial production away from the surface of the Earth and into orbit, creating a system in which the standards of living for both orbiting and Earth-bound populations would continually increase. The only hindrance to O'Neill's planning was the space shuttle, then in development. But O'Neill chose to believe that the shuttle would work as claimed, as a "well-founded prospect of low cost and routine operation" (Heppenheimer 1999, 435). And, with that assumption in place, a new frontier waited. As O'Neill himself put it, "what do we need for the exponential growth of wealth? Three things: energy, land area, and materials" (O'Neill 1976b, 35).

Yet, before O'Neill could create his own version of a new frontier, he first had to attempt to claim NASA's as his own. Applying for NASA's "group 6" astronaut-scientist program in 1966, he made it to the final selection before missing out. At the time, O'Neill was a recently tenured professor of physics at Princeton University. Known within the academy for the invention of a high-energy storage ring for particle accelerators, he was virtually unknown outside it (Dyson 1993).[3] Had he made it through the selection process at NASA his career would have been remarkably different: none of those eventually selected in 1967 flew into space until the space shuttle era and the advent of "mission specialist" positions in 1981. However, despite missing the selection for the astronaut corps, for the rest of his professional career as a physicist at Princeton University, and as a space advocate, the image of the frontier and the "conquest of space" propagated by NASA would influence O'Neill's own concept of the High Frontier and what he would eventually call the "humanization of space." Above all, however, the High Frontier would promote an image of a frontier lost in American myth that could be regained in the orbital colonies O'Neill was proposing for the very near future. Building from his expertise in physics and his failed attempt to join NASA, O'Neill synthesized what was an unoriginal set of ideas about the colonization of space in a way that fully resonated with an American nation eager to regain the benefits of some kind of frontier.

Having received his doctorate in physics from Cornell University in 1954, O'Neill became an instructor of physics at Princeton, a position designed to emphasize his work in the classroom rather than as a researcher. By 1956, however, his proposal for a magnetic storage assembly for particle accelerators allowed him to move into research as an assistant professor. (O'Neill 1956) That work, as physicist Freeman Dyson noted in O'Neill's obituary over three decades later, radically changed high-energy physics. "In two pages . . ." Dyson wrote, O'Neill "laid down the path that high-energy physics followed for the subsequent 36 years" (1993, 97). It would be his major contribution to high-energy physics, but the fundamentals of the process would be important in his work on space colonies later. And, by 1965, based on the particle ring storage device, O'Neill had tenure at Princeton, despite comparatively few major publications, and by the time he tried to enter the astronaut corps, he was a full professor of physics there, where he would remain for his entire academic career until his retirement in 1985.

Perhaps not altogether surprisingly since he was initially hired as a classroom Instructor, O'Neill took his duties as a teacher seriously and during the summer of 1969 revamped the organization the yearlong introduction course from the ground up. By today's classroom standards, the changes he implemented seem minor: replacing blackboard use with an overhead projector so he could face the students and using "learning guides" rather than traditional homework assignments, in an effort to motivate self-directed learning (O'Neill 1976b, 276). Perhaps most significantly, however, O'Neill tried to turn the normal introduction to physics topics toward the practical and the interesting. As far as O'Neill was concerned, in the fall of 1969 what could be more interesting to a group of physics students than space? It was certainly interesting to their professor.

While not selected for astronaut training, O'Neill pursued his interest in space, in part, through his teaching. It was in that teaching preparation that O'Neill had what writer Michael Michaud called his "'Big idea' of potentially historic proportions" (1986, 60). O'Neill's "big idea" was fueled both by his thwarted hope for astronaut selection and by a desire to engage Princeton students during the tumultuous 1960s. The Apollo program, as far as O'Neill was concerned, "offered many possibilities for illustration in a freshman physics course" (1976b, 277). Each of the major topics usually discussed in Physics 103–104 during the year—force, energy, momentum, thermodynamics, celestial mechanics, and electromagnetism—could be exemplified through the Apollo program. And O'Neill hoped that using those examples would provide a unifying framework for the year while also maintaining his students' interest.

The students O'Neill expected were "prospective physics and math majors, engineers, a sprinkling of other would-be scientists, and an occasional brave premed" (O'Neill 1976b, 276). O'Neill worried that his students required a new kind of perspective from the class, one that would mesh with their ideals for social change. Describing the period, O'Neill wrote, "the horrors of the war in Southeast Asia had provoked a revulsion against authority and against technology on American university campuses" (1976b, 276). No longer accepting of scientific and technological research for its own sake, Princeton students, it seemed to O'Neill, largely distrusted technology and science. As he put it during an interview with *Whole Earth Catalog* founder Stuart Brand in 1975, "this was the peak time of disenchantment with anything in science and engineering . . . And I felt that, despite the bad times, improvements in the human condition could be reached by using science and engineering in the right way, as opposed to the wrong ways" (Brand 1977, 22). In an attempt to direct the discussion toward what O'Neill considered the "right ways" and the practical, real-world social issues that might appeal to both his student's idealism and encourage an interest in science, O'Neill asked a select group of his students in that fall of 1969 a question that would alter his own career path dramatically. And, exactly as O'Neill hoped, it captured his students' attention. The question, planned during the summer of 1969, was a way to direct student interest toward real-world social answers through the tools of a physicist. Directed toward a small section of the 180 students in Physics 103, and initially designed by O'Neill as a question that would occupy them for, at most, a week or two, it would turn into a problem that consumed his students for the entire semester. The question he asked was simple, and in retrospect rather obvious: "Is," O'Neill asked a small selection of his freshman, "a planetary surface the right place for an expanding technological civilization?" (O'Neill 1976b, 279). Fueled by the Apollo lunar missions and O'Neill's unsuccessful attempt to join NASA, it turned out to be a question that worked in the classroom if not, initially, in his professional life.

O'NEILL'S PUBLISH-OR-PERISH IDEA

O'Neill's now seminal essay, "The Colonization of Space," was sent off to numerous journals, including the refereed journal *Science*, the popular magazine *The Atlantic*, and *Scientific American* (O'Neill 1974). Although *Science* and *Scientific American* were publications in which O'Neill had previously placed his work, reviewers at each rejected it, as did the editors at *The Atlantic*. According to O'Neill, an editor at *The Atlantic*,

> [w]as quite interested and he wrote back and asked an enormous number of questions, and then I expanded the article, answered all the questions, and sent it back again . . . Long months went by and finally he rejected it and said "it's not that I'm not interested, it's just that I have this feeling that I've asked ten questions and you've given me ten answers and now that suggests a hundred more questions, and it's not going to converge somehow. Better that I not do anything about it."
>
> (Brand 1977, 23)

Originally written during the height of the Apollo program, O'Neill was at a loss to explain why his ideas could not find a home. By his own account he rewrote the article multiple times, attempting to focus it for each publication's needs (Brand 1977, 23). His work in the classroom and during seminars with select students had convinced O'Neill of the importance of his ideas, but, as he noted later, "it was a long and very frustrating period before anything really came out on it . . . there was a period of four years during which I was trying to get it published and couldn't" (Brand 1977, 23–24). During the long waiting periods between sending off his manuscript and waiting for yet another rejection O'Neill was forced to become his own public relations spokesman. Beginning in 1972, he began giving talks at numerous colleges and universities along the East Coast or in California, usually to receptive undergraduate audiences. With what he described as "fairly primitive graphics," O'Neill honed his message to one of near-utopian simplicity (Brand 1977, 24). When *Physics Today* accepted the article in the spring of 1974 it was "about the 6th or 7th draft" O'Neill had submitted to various publications (Brand 1977, 24).

Physics Today, published by the American Institute of Physics since 1948, has a limited and mostly professional readership. But it is by no means a journal of original research. Physicists referee submissions, but the purpose of the publication has long been to provide information on trends within various areas of physics research to readers working in other subspecialties. It thus publishes essays without overly technical language unique to any special foci within the field. O'Neill's years of reworking drafts for journals and magazines such as *Science*, *Scientific American*, and *The Atlantic*, had produced an essay filled, primarily, with mathematical foundations for his ideas on space colonization combined with limited statements about the social benefits of those ideas.

But it would be the social benefits, not the math, which made O'Neill's "The Colonization of Space" an immediate success. Published in 1974, by 1975 the reception of the High Frontier concept would belie its initial publishing difficulties. In May 1975 O'Neill would host the Second Princeton University Conference on Space Manufacturing, itself leading to a front-page story in the *New York Times* (Sullivan 1974). By July of that year O'Neill would testify before the U.S. House of Representatives Subcommittee on Space Science and Applications of the House Committee on Science and Technology (cf. *Space Colonies* 1977, 12–21). In January 1976 O'Neill would offer similar testimony to the Senate Subcommittee on Aerospace Technology and National Needs of the Senate Committee of Aeronautical and Space Sciences (appendix 2 of O'Neill 1976b, 313–326). His book, *The High Frontier*, would be published in 1977 and would further refine both the technological and social benefits he believed would be the consequence of his High Frontier. And those social benefits were, explicitly, another version of American frontier utopianism. One key to the success of Frederick Jackson Turner's own "Frontier Thesis" has long been considered his idea of the "social safety valve." For O'Neill the notion of the High Frontier was a safety valve not just socially, but on countless other issues in need of solutions in 1960s and 1970s America. But, like Turner before him, it was the concept of a "frontier" that made all his solutions to contemporary problems possible. Virtually none of the ideas in either the "Colonization" essay or *The High Frontier* were original. But, as Howard E. McCurdy has noted, "[t]he degree to which they captured the public imagination was" (1997, 152).

Indeed, the lack of originality in O'Neill's High Frontier cannot be read as an attack on his work. The success of the work, however, does imply the importance of the frontier mythology O'Neill espoused. As McCurdy noted, in detail,

> [t]he idea of large artificial colonies in space had been advanced previously by an assortment of writers, from the famous to the obscure. At the beginning of the century Russian space pioneer Konstantin Tsiolkovskii had envisioned dwellings in space that could house millions of people. British scientist J.D. Bernal advanced a similar concept in 1929, and Arthur C. Clarke helped popularize the idea in his 1954 children's novel *Islands in the Sky*.
>
> (1997, 152)[4]

The more technical specifics of O'Neill's High Frontier all predate his presentation of them, including those that appear the most original. O'Neill's magnetic accelerator, or "mass-driver," used to launch material from the surface of the moon is in part a significantly larger revised version of his 1956 *Science* article for the magnetic storage of subatomic particles while also being nearly identical to a device described in Robert A. Heinlein's 1966 best-selling science fiction novel *The Moon Is a Harsh Mistress* (Heinlein 1966; O'Neill 1976a). The concept of solar power satellites had been proposed by Peter E. Glaser in 1968 and, in 1976, he had refined the concept for the American Astronautical Society in much greater detail than O'Neill (Glaser 1968, 1976). Science fiction scholar Gary Westfahl, in his examination of the fictional idea of space colonies, repeatedly shows that literally everything O'Neill proposes was proposed by others in both fiction and non-fiction, earlier, and in more detail (Westfahl 1996, 36–38). What O'Neill did, as Michael A. G. Michaud describes it, was "put together a grand synthesis of several ideas, proposing a complete system rather than a single large technological project" (1986, 65). While not original in any way, O'Neill's High Frontier manages to combine several lines of thought in American culture and add to them with a very liberal dose of the myth of a never-ending frontier. It was similar, therefore, to what Teddy Roosevelt famously conceded to Frederick Jackson Turner: O'Neill "put into shape a good deal of thought that has been floating around rather loosely" (cited in Billington 1973, 130). American political scientist Terrence Ball has described it somewhat more specifically. With his technologically advanced orbiting colonies that offer a "controlled environment, seasons, sunlight, and water supply [that] make the space station an ideal environment for farming," O'Neill has advanced, "in effect . . . and apparently quite unconsciously . . . The Garden in the Machine" (Ball 1995, 288). And this new version of a New World Garden would, conveniently, solve all the problems O'Neill saw currently facing the world.

O'Neill set forth his intent quickly in *The High Frontier* and considered his reasoning practical and straightforward as to why the effort at space colonization should be expended. The immediate need of, "solving the energy crisis which we face here on Earth," could be solved through his High Frontier, as could

> the slightly longer-term problem of population size and Earth's capacity to support it; [and] finally to a nonmaterial problem, compelling but not to be reckoned in dollars: the opportunity for increased human options and diversity of development.
>
> (O'Neill 1976b, 14)

And O'Neill firmly believed that each of these issues—energy, population, and a "diversity of development"—came through his High Frontier because that idea came with "unlimited new lands, to provide living space of higher quality than that now possessed by most of the human race" (1976b, 33). Indeed, O'Neill's rhetoric is an expansive restatement of Turner's "Frontier Thesis," placing the High Frontier into a utopian future rather than a mythical past.

The rhetoric in *The High Frontier* was based, in part, on American science fiction and, in part, on the frontier ideology that had inspired it. It was not just the failed astronaut who proposed the High Frontier; he was also a self-claimed, "reader of science fiction in childhood," whose wife noted in a 1977 *People* magazine interview, knew every *Star Trek* episode "by heart" (O'Neill 1976b, 60; Klein 1977, 24). Containing the technical specifications of the "Colonization" essay, updated with some new ideas as well, much of the book used a series of fictionalized "letters home" from colonists in an orbiting habitat. These letters bear a striking resemblance, rhetorically, to forms of utopian and science fiction literature, espousing as they did the wonders of O'Neill's vision. *The High Frontier* employed a rhetoric of desire for room to expand, and promised unlimited potential for those willing to homestead on such a frontier. They capture, at once, the science fiction conceit Tom Moylan describes as a place often "better than the present" that then takes that place and "spins them around within the one and only 'paradise' that is allowed to exist" (2000, 29). And, perhaps counterintuitively, O'Neill's letters also offer the tradition of American utopias that have had a long tradition of pastoralism while combining it with the significant utopian strain that promoted technology. A surprising number of utopias written by Americans in the nineteenth century, according to Jean Pfaelzer, "fulfilled the hopeful tendencies of democracy, technology, and current ideas of social and economic reform" (1984, 19). Indeed, as Howard P. Segal has described the tendency in American utopias with a specific technological theme, the writers of such works "actually believed that advancing technology would be the key to turning the impossible into the possible and even the probable" (1985, 2). The High Frontier does precisely this, while also capturing a frontier in terms Americans had thought lost to history. Anxiety over such lost frontiers had long been central to the American conception of itself in the twentieth century, and O'Neill makes it clear that his orbiting space colonies could replace that long-lost frontier. Images of the frontier are front and center in each presentation of the possibilities inherent in his High Frontier:

> The era in which virgin land of good quality in a good climate was available for settlement is long past; the United States is a relatively uncrowded country by world standards, but already our fastest growth is in regions (Arizona, New Mexico, and other desert areas) which would not attract large numbers of people if there were no air conditioning.
>
> (O'Neill 1976b, 36)

What the orbiting space colonies will provide for their inhabitants is what O'Neill believed the long-lost American frontier provided for the original settlers of the continental west:

> [There are] unlimited new lands, to provide living space of higher quality than that now possessed by most of the human race [and] an unlimited materials source, available without stealing, or killing, or polluting.
>
> (O'Neill 1976b, 33)

However, O'Neill's promised future and boundless frontier also overcome those aspects of the American frontier that were, for a socially conscious university professor, problematic: "There will be no 'hostile Indians' and there will be plenty of food" (1976b, 220). In the High Frontier O'Neill combined ideas from throughout popular culture, science fiction, and space sciences. The end result is a proposal with a kind of "something for everyone" attitude. As political scholar Howard E. McCurdy puts it, O'Neill's vision could be

> embraced by visionaries who wanted to pioneer space, environmentalists concerned about overpopulation and dwindling resources, futurists who saw technology as the solution to human problems of industrial civilization, and a variety of space-age groupies in occasional need of psychiatric help.
>
> (1997, 152–153)

NOT TIN CANS BUT *ISLANDS* IN SPACE

This main concept O'Neill proposed in both his "Colonization" essay and *The High Frontier* was to colonize space. But, the vision of those colonies was far more than weightless, claustrophobic environments. O'Neill called them "Islands in Space," and they were that and more. Taking the shape of large spheres, rings, or cylinders, they would occupy gravitationally stable locations in the Earth–Moon system. Those five stable locations, named after the French Prussian mathematician Joseph-Louis Lagrange, who originally hypothesized them in 1772, came to be called Lagrange Points, and O'Neill would hope to place his first colony, Island One, at the Lagrange-Five (L5) location, along the moon's orbit around the Earth.

Island One was to be "small," perhaps housing only ten thousand people. O'Neill outlined the technical specifications in his "colonization" essay: it could be built with a construction force of two thousand people and weighing an approximate five hundred thousand tons (1974, 38). Island One's population would then build a larger Island Two, also at the stable L5 location. But, it would be later, with the building of what O'Neill called Island Three by the first two colonies that the true frontier gains would begin. The building of Island One should begin almost immediately in O'Neill's view, thus placing the building of Island Three in the early to mid-2000s, and well "within the limits of present [early 1970s] technology" (1976b, 64). Island Three would be constructed as a cylinder, the materials for which would be processed from moon rock and shipped to the L5 location. O'Neill's vision saw Island Three having, "a diameter of four miles, a length of twenty miles, and a total land area of five hundred square miles, supporting a population of several million people" (1976b, 64–65).

Such a vision was achievable, in O'Neill's view, because the technology already existed and the benefits far outweighed the cost. He had made his most grandiose claims right at the beginning of his "Colonization" essay in 1974 with simple bullet points:

- We can colonize space, and do so without robbing anyone and without polluting anything.
- If work is begun soon, nearly all our industrial activity could be moved away from Earth's fragile biosphere within less than a century from now.
- The technical imperatives of this kind of migration of people and industry into space are likely to encourage self-sufficiency, small-scale governmental units, cultural diversity and a high degree of independence.
- The ultimate size limit for the human race on the newly available frontier is at least 20,000 times its present value. (O'Neill 1974, 32)

Yet not content to hypothesize simple arguments for why and how easy the colonization of his High Frontier would be, O'Neill turned, near the end of that essay, to a more grand proclamation:

> It is important to realize the enormous power of the space-colonization technique. If we begin to use it soon enough, and if we employ it wisely, at least five of the most serious problems now facing the world can be

> solved without recourse to repression: bringing every human being up to a living standard now enjoyed only by the most fortunate; protecting the biosphere from damage caused by transportation and industrial pollution; finding high quality living space for a world population that is doubling every 35 years; finding clean, practical energy sources; preventing overload of Earth's heat balance.
>
> (1974, 36)

Each of these claims represented a significant argument against a number of books published in the 1960s, including works on energy, population, poverty, and the environment. And, in the details provided by O'Neill the solution is always the colonization of space and the resulting orbital Utopia.

O'NEILL'S HIGH FRONTIER AS UTOPIAN SPACE

Gerard K. O'Neill spends much time in *The High Frontier* claiming that his vision of orbiting space colonies is not "Utopian." It goes far beyond such earth-bound concepts because, for O'Neill, the concept of utopia is built on both "restricted social ideas and restricted technology" (1976b, 236). For O'Neill a utopia is essentially a negative thing: at once impractical and, by necessity, limiting of social freedoms. However, the claim that the High Frontier is not utopian belies the very specific goal O'Neill lays out, or at least believes his L5 colonies will accomplish, and that is to create a system in which the problems of his current society are lessened, if not wholly eliminated. Writing in 1977 T.A. Heppenheimer's *Colonies in Space* is largely a summation and extrapolation of O'Neill's ideas. Heppenheimer, who took part in the NASA summer study cited at the beginning of this chapter, and who holds a Ph.D. in aeronautical engineering, would eventually become a well-known historian of flight and spaceflight—and is frequently cited here in the chapter on NASA. Heppenheimer, if anything, offers O'Neill's logic in a better fashion than does O'Neill, himself. Heppenheimer suggests that the High Frontier cannot be utopian because, significantly, it is a *frontier*:

> The colony will not be a paradise. It will be comfortable and attractive, but it will still be a frontier community working hard to build power satellites. Its population will be too small to sustain a major university, a large medical center, or the profusion of specialized shops and stores which are found in all large cities . . . though the colony would be largely self-sufficient for its major needs, there are items for everyday life which it would have to import . . . As with the early colonists of America, these items would come by ship from the mother country.
>
> (1977, 143)

Yet, despite Heppenheimer's statement that the colonies would be too much a frontier society to be a paradise, O'Neill's argument that these frontier communities would produce an end to all the problems currently confronting the Earth still implies that the High Frontier is a utopian vision. It is a model of the future that is utopian, in the sense meant by utopian scholar Ruth Levitas when she defines the term, very generally as, "the construction of imaginary worlds, free from the difficulties that beset us in reality" (1990, 1). What O'Neill accomplishes rhetorically through his High Frontier is precisely the purpose of the more traditional literary utopia. While O'Neill can simultaneously claim that he is both writing about a more perfect future and not writing an actual utopia, he is doing exactly what a utopian novel undertakes: he is considering the present.

O'Neill makes it clear that he "has no desire to influence or direct in any way, even if I could, the social organization and details of life in the [space] communities" (1976b, 235). Yet, what he finally presents is an image of an orbiting world that "we should hope that they will give added opportunity for that most elusive of human occupations, so fundamental as to be written into our Constitution: 'The pursuit of happiness'" (O'Neill 1976b, 273). It remains a utopian vision not because it provides an extensive plan, but for its desire to not provide such a plan. Where *Star Trek* offers a fully realized utopian vision of the twenty-third century, O'Neill's High Frontier merely offers the technical specifications for orbiting space colonies. But with those plans comes, in classic utopian fashion, "the desire for a different, better way of being" (Levitas 1990, 181).

More fully in the book-length *The High Frontier* than in his "Colonization" essay, O'Neill is concerned with solving the problems noted by his readers through a rhetorical strategy built on the tropes of both utopian literature and science fiction. As Tom Moylan has noted in his *Demand the Impossible: Science Fiction and the Utopian Imagination*, such a statement is central to the utopian message. "For as 'realistic' as science fiction and Utopian fiction may appear on the surface," Moylan notes,

> they are forms of the romance which are meditations upon the deep conflicts in the historical present that are displaced onto the terrain of an other-worldly locus so that the reader, consciously or unconsciously, can see her or his society and its contradictions in a fresh and perhaps motivating light.
>
> (1986, 32)

Moylan's own definition of *utopia*, which is drawn extensively from that of Frederic Jameson, defines the term around its ability to posit both general and specific solutions to the contemporary problems of the reader. A truly Utopian work for Moylan is a work that addresses, for the majority of its planned readers, ways to understand their own problems and envision solutions. In both his "Colonization" essay and *The High Frontier*, O'Neill does exactly what a utopian novelist would undertake: a detailed examination of the social and physical structure of an imagined society. Moylan notes that what is classified in a traditional novel, as a "mere background setting" is, for a utopian text, "the key element of the text." Normally in such narratives, the background setting is an extended examination of "the society projected in such a complete manner as to include everything from political and economic structures to the practices and rituals of daily life" (Moylan 1986, 36). For a work of literature to be truly utopian, then, it must consider such things as social structure and justice systems. Yet, while O'Neill repeatedly claims this is exactly what he is *not* doing, the ramifications of his technical designs for the High Frontier continually reveal otherwise.

Thus, O'Neill claims he is concerned with something far more specific in his High Frontier: the actual physical structure, the engineering rules of its assembly, and the strict scientific examination of its source of raw and finished materials and energy. The "political and economic structures" that is central to Moylan's definition will, with the correct orbital velocity and stability, follow as a matter of course. This is why O'Neill's High Frontier can both project the traditional, mythological needs of the American frontier onto a new canvas and, quite specifically, address the problems O'Neill envisions his readers would consider centrally important. Through his use of both utopian themes and, eventually, the rhetoric of science fiction, O'Neill is able to provide exquisite technical details of his orbiting habitats and, without worrying about the details too specifically, claim that such an environment will provide solutions to his readers' contemporary concerns. And while he will claim these ideas are most certainly not utopian, his rhetoric betrays that claim again and again.

O'NEILL'S "LETTERS FROM SPACE" AND THE TRAVEL NARRATIVE TRADITION

For all of O'Neill's protestations to the contrary, when in the opening pages of his *The High Frontier* he begins with "A Letter from Space" it is difficult not to consider his narrative to be utopian. Writing from "as early as the 1990s," his first letter home—there will be three in *The High Frontier*—conforms almost too well to the tradition of utopian fiction and the tradition of travel narratives in American history (O'Neill 1976b, 3). His letter writers, "Edward and Jenny," not only describe life in the orbiting High Frontier colonies in detail but also offer a description of the colony, itself:

> All the habitats are variations of basic sphere, cylinder or ring shapes. We live in Bernal Alpha, a sphere about five hundred meters in diameter, with a circumference inside, at its "equator" of nearly a mile . . . The land forms a big curving valley, rising from the equator to 45° "lines of latitude" on each side. The land area is mainly in the form of low-rise, terraced apartments, shopping walkways, and small parks . . . we like to preserve most of our land area for grass and parks. Our sunshine comes in at an angle near 45°, rather like midmorning or midafternoon

> on Earth . . . Our apartment is about the same size as our old house on Earth, and it has a garden.
>
> (O'Neill 1976b, 6–7)

Any familiarity with Thomas More's *Utopia* demands a comparison to that work's description of the Utopian city of Amaurot, which almost five centuries earlier is characterized in profoundly similar ways. While it is unlikely O'Neill intended such an obvious similarity, the fact of that similarity is key to understanding O'Neill's High Frontier as utopian in scope:

> It [Amaurot] lies upon the side of a hill, or, rather a rising ground. Its figure is almost square, for from the one side of it, which shoots up almost to the top of the hill, it runs down, in a descent for two miles, to the river Anider . . . Their buildings are good, and are so uniform that a whole side of a street looks like one house . . . there lie gardens behind all their houses.
>
> (More 2008, 46)

For O'Neill his orbiting space colonies will, he continually notes, emphasize parks and gardens. But where Karen R. Jones and John Wills can note that "the park ideal has, as yet, proved incapable of furnishing the human race with paradise or utopia," O'Neill's vision offers exactly such a utopia (2005, 175). O'Neill's fictional letters home continually refer both to the frontier and to the parks and gardens people will build there as they totally control nature. And where Jones and Wills can effectively argue that parks "hardly represent total models on which to base entire societies," O'Neill seems to have no difficulty implying that they, indeed, do (2005, 176).

The tradition in utopian narratives of the "letter home" or the "recollections of travel" for the reader is among the most common tropes of all Utopian fiction. From this perspective the narrator—"the guide" in Northrop Frye's view—can offer explanations of "the structure of the society and thereby the significance of the behavior being observed" (Frye 1973, 26). Like More's *Utopia* then, O'Neill's "letters from space" offer the same literary convention of many utopias. However, if a direct comparison to any classical utopia were to be offered, it is likely Francis Bacon's *New Atlantis* (1915) that is O'Neill's direct literary ancestor. The *New Atlantis*, and O'Neill's High Frontier, both present images of a utopia in which "a society where science flourishes" makes the Utopia, itself, possible (Hutton 2002, 48). And, like Bacon's work, O'Neill often seems far more interested in the technology and science he is espousing and assumes the utopian benefits will simply accrue as a matter of course. Yet, for all of the utopian narrative conventions that O'Neill utilizes it is important to realize that such traditions are also part of other significant and lengthy traditions leading to not just utopian literature but to the travel narrative of European colonialism and science fiction as well. Each of O'Neill's three letters contains the narrative tropes of each of these traditions and, to a lesser extent, the tropes of the newer postcolonial category of "travel literature." Writing on that latter, field sociologist Chris Rojek had offered a description that applies equally to utopian, science fiction, travel and tourist literature, and O'Neill's fictional "letters from space." Rojek notes that the purpose of such narratives is to provide, "as a social category 'the extraordinary place' [which] spontaneously invites speculation, reverie, mind-voyaging and a variety of other acts of imagination" (1997, 52). O'Neill's fictional letters accomplish each of these things, often in surprising detail. In the tradition of travel narratives in colonial America, however, where the location of travel, as Wendy Martin has described them, is "a land that was depicted by some as a howling wilderness and by others as a new Eden," O'Neill comes down firmly on the side of new Eden, all the while denying his colonies as such (1994, vii). Martin, an American historian, has also described the narrators of such colonial documents with a specific narrative arc in mind: "so the secular traveller responds to physical challenges as an opportunity to demonstrate superior character." For O'Neill's fictional travel narratives, however, this is not quite the case because there is little need for heroism such is the environment provided in the orbiting islands. Thus, there is no need for what British critic Richard Kerridge claims is a requirement: a narrator "played by the traveller as heroic venturer, the person whose very daring in entering these places emphasises their difference and impenetrability, while giving the reader a sensation of sudden closeness to the wild and foreign" (1999, 165). O'Neill, instead, goes to great lengths to suggest that his High Frontier is neither wild nor foreign—no "howling wilderness" here—but would be merely commonplace. By 1981 he would ask, rhetorically, "you might then ask whether ordinary people in large numbers will ever find it easy to travel into space," and he answers, after some very minor math, with the idea that with a second-generation "1990s vintage" space shuttle and other "shuttle-derived vehicles" there "would be quite enough to provide transport . . . for several hundred million people per year" (O'Neill 1981, 73–74). No "heroic venturer" was needed—everyone could immigrate to the High Frontier. O'Neill would offer the shuttle as "an excellent launch vehicle even for an ambitious, large-scale program of space manufacturing," and, with *that* shuttle (which was clearly a more optimistic image than the one NASA eventually launched), offer a narrative of everyday Americans living and working in orbiting habitats and sending letters home about their quality of life (O'Neill 1978, 32).

But, if not fully travel literature, the High Frontier also can lay claim to those qualities that define science fiction. As American writer Robert A. Heinlein attempted to define that genre in the late 1950s, O'Neill's High Frontier is clearly "realistic speculation about possible future events, based solidly on adequate knowledge of the real world, past and present, and on a thorough understanding of the nature and significance of the scientific method" (1977, 4). However, while not fully a fictional story, *The High*

Frontier is, in O'Neill's own words, a work contending that "we now have the technological ability to set up large human communities in space: communities in which manufacturing, farming, and all other human activities could be carried out" (1976b, 13). The entire book presents, then, what O'Neill hopes are "realistic speculations" about "possible future events," and as a professional physicist his "adequate knowledge" of the "scientific method" cannot be doubted. However, perhaps, his knowledge of "the real world, past and present" is somewhat narrowly focused and, certainly, is not the knowledge of a trained historian, sociologist, or, indeed, any social scientist. Nonetheless, the High Frontier does offer a narrative with many of the conventions of science fiction.

Yet, both the narratives of the three "letters from space" and the eventual benefits of his High Frontier mark O'Neill's concept as, most closely, utopian in nature. While it remains impossible to fully disengage O'Neill's writing from either science fiction or travel narratives—and each of these areas of inquiry offers much insight into his work—the final conclusion remains that, at heart, the High Frontier as a concept and *The High Frontier*, itself, remains essentially utopian works. To examine this contention it remains most fruitful to follow the narratives of O'Neill's "letters from space."

His first letter back to Earth, referenced earlier, offers details on the political and legal organization of Bernal Alpha: "there are almost as many different kinds of local government as there are national groups within the colonies; ours happens to be a town-meeting style." Also well described is a general image of the wonderful "homesteading" experience in the colony: "because of our unlimited cheap energy we don't have pollution here"; "We . . . never need to use insecticide sprays"; and "Jenny is an enthusiastic gardener" (O'Neill 1976b, 8, 10). But, in the tradition of colonialist travel narratives, it posits the extension of the life of the "home country" to a "new territory." As Mary Louise Pratt has described early colonial writing in the sixteenth to the eighteenth centuries as narratives in which "the systems of nature [that] were projected within European borders" were also narratively extended "beyond them," O'Neill's colonists spend an inordinate amount of time describing those parts of "nature" they have brought with them into space—and, in the letters there are all to frequent mentions of that "nature" in the form of "gardens" (1992, 35). But where Pratt sees the invention of the Linnaean classification system as an enabler of European colonialism, O'Neill merely transports both those Enlightenment scientists, and their backyard gardens, into space.

The second letter, far later in the text, adds to this with the initial views of new colonists. The writers, "Jennie" and her husband, exclaim "after Michigan in January we're happy to soak up the sunshine for awhile," and describe the environment aboard "Island One" as "really . . . sort of a vacation atmosphere" (O'Neill 1976b, 209). Encouraging their friends to journey to the new colony Jennie expounds on the quality of food, the gentle weather, the four-day workweeks, and ends by proclaiming to her readers, friends back on Earth "Peggy and Arthur," that should they come for a "second honeymoon," "you may never want to go back. Now that we've found what it's like, I can tell you it's going to be a lot harder for us to leave!" (O'Neill 1976b, 215).

By the third and final fictional letter O'Neill provides the colonists are so happy that the only desire they have remaining is to venture further into the High Frontier—in this case, to the asteroid belt between Mars and Jupiter. A "Dad and Mom" begin dictating a letter to their young son, Stephen, as they prepare to move to the asteroids, so that he will understand their decision to relocate from Earth orbit. Following their entire planning stage as well as their voyage they end with encouragement about their new frontier, exclaiming: "We've got a whole world to build here, Stephen, so grow up fast and get in on the construction!" (O'Neill 1976b, 245). Perhaps even more indicative of precisely what kind of mythic tropes O'Neill is evoking, however, the parents of young Stephen describe how they and a number of their friends—all apparently nuclear family units—plan on building separate ships and travel to the asteroids:

> Out here, though [in Earth orbit], we're in much better shape to go voyaging on our own. Our spacecraft never has to take big forces, and our engine can be small . . . I've been building it in the marina, near the docking ports of Gamma, and we've checked the welds with x-ray equipment that we've borrowed from the plant. Alongside [our ship], at the marina, there are four more of the same kind . . . the others are going with us, in a "wagon train" of five craft, so that if any of us runs into trouble either before or after we arrive, there'll be help near at hand.
>
> (O'Neill 1976b, 239–240)

If *Star Trek* was intended as a "*Wagon Train* to the stars" with the military-industrial complex of Starfleet, O'Neill's "Wagon Train" is far more of a Jeffersonian yeoman family affair. In his letters home O'Neill insistently proclaims that "space habitats are exciting places to be," and this excitement seems to be tied up in the stability of married couples there—each letter is from a couple to either another couple or their child—and all are about the large number of families, with homes *and gardens*, in all the colonies. The consistent control over nature implicit in such statements both highlights the themes O'Neill is really espousing—science and technology—or serves as a stark reminder of the greater implications of the High Frontier. Palestinian American scholar Edward Said once proclaimed about his concept of orientalism that it "embodies a systematic discipline of *accumulation*. And far from being exclusively an intellectual or theoretical feature, it made Orientalism fatally tend towards the systematic accumulation of human beings and territories" (2003, 123; original emphasis).[5] Said's line of reasoning has since been applied throughout much of postcolonial literature, and it certainly applied to the colonial experience in the Americas that O'Neill seems to draw from rhetorically. As Stephen Greenblatt has claimed, "the whole experience of Europeans in America was shaped . . . by a particularly intense dream of possession," and in a way Greenblatt's "dream of possession" or Said's more formalized "discipline of accumulation" can also be applied to the High Frontier (1991, 121).

But O'Neill, in one sense, is able to present a "systematic discipline of accumulation" *and* a "dream of possession" and to not worry about the ethical implications. Like Captain John Smith declaring Virginia to be a place that "heaven and earth never agreed better to frame a place for means of habitation," O'Neill does present the High Frontier as a sort of paradise with gardens and weightless flight and nearly unlimited energy in his "letters from space" (Smith 1973, 2).[6] But at the same time the High Frontier has no "other people" to conquer and colonies. Again, Edward Said has offered a postcolonial perspective on the American past that O'Neill is offering up in his High Frontier. Said has noted that, for Europeans and eventually for Americans, "there were native peoples to be dominated, variously exterminated, variously dislodged," but the clue to O'Neill's Utopian pretentions is that, on the High Frontier, no such people exist until they have moved there (1993, 7). And, there are no new parts of the map to color in the shade of one's own empire. Instead, the logic of this empire can be used in O'Neill's vision for the "accumulation of human beings and territories" at little or no cost in the vast emptiness of space—either ethically or economically. And, finally, the utopian vision O'Neill provides is shown as much in the contemporary quandaries he claims to solve with his High Frontier as it is in the imaginary societies he hints at.

MODELS OF DOOM: TRYING TO NOT WORRY SO MUCH ABOUT THE FIN DE SIÈCLE

There is a significant literature from the 1960s and early 1970s that posit a truly apocalyptic future for humanity. The "Commission for the Year 2000," founded in 1965 by the American Academy of Arts and Sciences and chaired by sociologist Daniel Bell, aimed for social scientific analysis of trends into the future, eventually publishing multiple volumes of "working papers" on the topic (Bell and Graubard 1968). More academic oriented works followed, such as Bell's 1971 *The Coming Post-Industrial Society* (1973) and Herman Kahn and Anthony J. Wiener's *The Year 2000* (for which Bell provided an introduction; 1967). These academically oriented works all presaged an emerging worry about "the future" generally. However, it is the more popular works of the period that firmly establish a general cultural concern over notions of "the future" that mirror the anxiety over Turner's closing of "the frontier" only eighty years before. In novels and films concerns about a coming disruption of the world system abound in works such as *On the Beach* by Nevil Shute (1957); *Make Room! Make Room!* by Harry Harrison (1966), from which the film *Soylent Green* (Fleischer 1973) was developed; and Stanley Kubrick's *Dr. Strangelove* (1964). The turning point for the American myth of the frontier, and thus the emergence of a new dystopian genre, is the 1950s. It is not surprising that the films of the period, as Susan A. George has claimed, "[feature] Earth and its people as the new frontier for a variety of usually unfriendly settlers, typically labelled 'invaders'" (2000, 77). While fictional filmed apocalypses dominate science fiction films of the 1950s, it will be the 1960s that see such themes migrate to nonfiction bookshelves.

It is in those nonfiction works of an imagined apocalyptic future that more fully inform a reading of O'Neill's "Colonization" essay and *The High Frontier*. Dominating best-seller lists, book clubs, and talk shows of the period, Rachel Carson's *Silent Spring* (1999) and Paul Ehrlich's *The Population Bomb* (1968) each argued from its own perspective that the future, lacking in frontiers of any kind, looked dim. In 1972, in a more academic format, The Club of Rome's *The Limits to Growth* (1972) brought together all of the apocalyptic visions of popular and academic culture to present visions of the future that were designed to be frightening to their readers. Through their status as "nonfiction" each of these books offered what is intended as a realistic and reasonable presentation of future problems. In addition, each of those apocalyptic futures is, directly or indirectly, "solved" by O'Neill's High Frontier. The four bullet points at the beginning of O'Neill's "Colonization" essay are responses, directly or indirectly, to these best-selling nonfiction works of apocalyptic worry. De Witt Douglas Kilgore, writing specifically on O'Neill's response to *The Limits to Growth*, offers an analysis that applies to each of the works under consideration here inasmuch as they are all books that are

> a critique of the ideology of progress in so far as that ideology rests on a thesis of endless growth and limitless accumulation. [They] also call . . . into question the primacy of the industrial revolution as the last, determinant stage in the economic and political evolution of humankind.
>
> (2003, 156)

A comparison to the problems detailed in such works as *The Limits to Growth*, and how O'Neill's utopian vision offers a solution to those problems can further reveal the frontier utopianism of the High Frontier. Additionally, since the rhetorical strategy employed by O'Neill is a significant part of his vision, a comparison to more dystopian works from the period that will employ similar rhetorical tropes is paramount here.

British economist Christopher Freeman, in a brief overview of a larger critique of the Club of Rome's *The Limits to Growth*, characterized it as "Malthus with a Computer" in a 1973 volume titled *Thinking about the*

Future (1973). But it is perhaps the American title of that same work, *Models of Doom*, which best describes a small genre that emerged, forcefully, with Rachel Carson's *Silent Spring* and reached maturity with Paul Ehrlich's *The Population Bomb* and similar works. Examples could easily include Alvin Toffler's 1970 book *Future Shock*, hypothesizing the distressing psychological results of "hyper-industrialization," or Howard J. Ruff's 1974 work *Famine and Survival in America*, advising readers to store a year's worth of food in preparation for a coming economic collapse. While *Models of Doom* was a critique of *The Limits to Growth* specifically, its conclusions easily apply to that genre of both academic and popular works of the 1960s and 1970s that foretell a calamitous, if not apocalyptic, future for humanity. The Models of Doom stories were a kind of dystopian nonfiction in which authors such as Carson, Ehrlich, and others cast themselves as prophets of a dismal future, setting themselves up, in the words of another contributor to *Models of Doom*, as "doomsday authors" (Page 1973, 172).

Silent Spring, *The Population Bomb*, and *The Limits to Growth* all present a future that is fully dystopian. Furthermore, each work presents its dismal future in what Tom Moylan has characterized as the model of a classical dystopia, in which, the "critical logic" of such narratives, "is a simplifying one" (2003, 136). Moylan suggests that the classical dystopian fear is the totalizing power of the state, but in the dystopias O'Neill reacts to that totalizing power remains human growth in a world without room to expand, or, put more succinctly, a lack of new frontiers. Two major analyses of O'Neill's High Frontier, Michael A. G. Michaud's *Reaching for The High Frontier* and De Witt Douglas Kilgore's *Astrofuturism*, each place the concept in opposition to *The Limits to Growth*. Michaud calls *The Limits to Growth* a "direct challenge to Western industrial and technological optimism" that O'Neill was trying to recapture in his High Frontier (1986, 58). Kilgore, noting that the thesis of *The Limits to Growth*, "created a firestorm in academic, industrial, and government circles" that could be answered by O'Neill's attempt to recapture the "endless growth and limitless accumulation" of a mythic frontier ideology (2003, 154, 156).

However, while both Michaud and Kilgore present useful readings of O'Neill's High Frontier, placing the idea in opposition to the entire Models of Doom genre is a far more effective strategy for understanding exactly what O'Neill was reacting against. While *The Limits to Growth* remains the most dramatic and effective totalizing work of Models of Doom genre, O'Neill's rhetoric dramatically differs from that work and contains much that is similar to the work of other doomsday authors such as Carson and Ehrlich. While the authors of *The Limits to Growth* specifically present their research as a culmination of an objective social scientific examination of available data, the other doomsday authors utilize the rhetoric of science fiction and dystopia in offering an image of a truly apocalyptic future. O'Neill, in response, uses the rhetoric of utopian science fiction to present a far more optimistic future. Thus, an examination of the Models of Doom genre of works from the period is a far more useful technique for understanding O'Neill's Utopian schema. In *Silent Spring*, Rachel Carson looks to the overuse of chemical pesticides as a way to criticize the entire paradigm of scientific progress and the control of nature. Paul Ehrlich extrapolates population growth into a truly catastrophic, Malthusian future in *The Population Bomb*. And both of these writers, to differing extents, present their concerns through a dystopian rhetoric designed to elicit an emotional as well as an intellectual response.

Thus, an examination of O'Neill's "Colonization" essay and his book-length restatement of that essay in *The High Frontier* require the context of placing it alongside the Models of Doom of the doomsday authors he is refuting. While that refutation is not always intentional, quite often it is. Furthermore, the narrative techniques used in *The High Frontier* is, strikingly, rhetorically quite similar to that found in *Silent Spring* and *The Population Bomb*. As a result, a comparison of O'Neill's "letters from space" to Carson's "fable for tomorrow" and Ehrlich's "three scenarios" can enlighten a cultural analysis of all of O'Neill's ideas. While *The Limits to Growth* does not utilize similar rhetoric, its placement as the culmination of the Models of Doom genre examined here makes it central to the context of O'Neill's work. The work of Carson, Ehrlich, and the Club of Rome by no means existed in an intellectual vacuum. Each work represents a significant cry of doom for the human population of the planet as no new frontiers remained for growth and existing areas could no longer be exploited through additional technological methods. Canadian geographers Pierre Desrochers and Christine Hoffbauer (2009) traced what they call "the post War Malthusian revival" to two 1948 best sellers. Fairfield Osborn's *Our Plundered Planet* (1948) and William Vogt's *Road to Survival* (1948) each provided an intellectual framework that Rachel Carson, Paul Ehrlich, and the Club of Rome would expand upon. In reacting to the dystopian rhetoric of the doomsday authors O'Neill employs a utopian rhetoric of his own. O'Neill offers a repudiation of the works of Carson, Ehrlich, and the Club of Rome and, to do so, builds more than simply castles in the air but a utopia in orbit.

IN SPACE NO ONE CAN HEAR *SILENT SPRING*

If *Our Plundered Planet* and *Road to Survival* represent the infancy of the Models of Doom then Rachel Carson's 1962 book *Silent Spring* is that genre coming to maturity. Carson was concerned the effects of chlorinated hydrocarbon pesticides on the environment, examined especially carefully through examples of severe unintended environmental consequences of overuse. This is, of course, both an intensively specific theme and one that Carson would draw out to quite general conclusions. Its specificity might make it appear an arbitrary choice for comparison to O'Neill's work. However, it represents both the first major work of the Models of Doom genre that would influence those volumes that followed and a work that O'Neill would directly contradict in representing his frontier utopia in orbit as a location where it would be possible to exert total control over the same "nature" that Carson argues should be protected. And, due to its initial popularity and its eventual influence, it has been called the first book of the modern environmental movement (Waddell 2000, 1). The conservative American news magazine *The National Review* (1999) named it one of the one hundred best nonfiction books of the twentieth century (at number seventy-eight) while *Discover* placed it at number sixteen of their "25 greatest Science Books of All Time" (Discover Magazine Editors 2006). Writing in 1970 science writer Frank Graham, Jr., called its publication, "one of those rare felicitous combinations of subject and author which produce a work of literary interest as well as an impact on a great social debate," and then, like many other critics, proceeded to compare it favorably with Harriet Beecher Stowe's *Uncle Tom's Cabin* (1970, 3). Others, however, have suggested that for inspiring the banning of pesticides *Silent Spring* has led to more deaths from tropical diseases such as malaria than were caused by Hitler and Stalin, combined, during and after the Second World War. This calculus makes Carson possibly the greatest mass murderer in history. On the occasion of Carson's one hundredth birthday, *Capitalism Magazine* ran a retrospective blaming her for "a million deaths per year," and *The National Review*, even after placing *Silent Spring* on their list of one hundred best nonfiction books of the century, blamed Carson for both millions of deaths and the birth of "alarmist environmental politics" (Lockitch 2007; Murray 2007).

Precisely how Rachel Carson came to write a book considered simultaneously both comparable to the antislavery classic *Uncle Tom's Cabin* and the paramount cause of unintentional mass murder is a story a woman inspired by the essential tensions between the "two cultures" of science and literature. *Silent Spring* is an excellent case study in the emergence and growth of the Models of Doom genre, and in the person of Rachel Carson the science writer we have a woman who hoped to do what all of the doomsday authors aspired to accomplish: make a case for a more careful relationship between the Earth and its human inhabitants. Where *Silent Spring* has been called a founding document of the modern environmental movement, it should also be seen as a founding text arguing against frontier ideology as well. As Carson's biographer, the American environmental historian Linda J. Lear, puts it, Rachel Carson desired to use science and her writing about the Earth to "widen the circle of people engaged in wondering at it, [so that] life would be safe from the evermore-destructive forms of human arrogance" (2000, 214). For Carson, simply, the frontier mentality of moving into a new environment and using science and technology to exploit it was, in the final analysis, exactly the method of "progress" she found objectionable, and *Silent Spring* would be her critique of such a worldview. Carson had previously written a trilogy of works on ocean life, including the 1951 best-seller *The Sea around Us*, which was serialized in multiple publications prior to being published as a full-length book. Winning the American National Book Award, it also established Carson's reputation. Furthermore, as literary scholar Mary A. McCay notes, in *The Sea around Us* offers a basic insight that would inform *Silent Spring*, namely, that it was "the human element that seemed always to upset the delicate balance of nature" (1993, 43).

Silent Spring argued that the overuse of modern pesticides, especially dichlorodiphenyltrichloroethane, or DDT, was both destroying the environment and leading to immediate and long-term negative health effects on humans as well. The "Silent Spring" of the title was Carson's thought experiment of a spring season in which birds, their populations devastated by the overuse of pesticides, were no longer present to sing. "On the mornings that had once throbbed with the dawn chorus of robins, catbirds, doves, jays, wrens, and scores of other bird voices," Carson wrote in the opening chapter of *Silent Spring*, "there was now no sound; only silence lay over the fields and woods and marsh" (1999, 22). It was those initial few pages, which Carson called "A Fable for Tomorrow," that provided the narrative impetus for the rest of the book—and bears comparison to O'Neill's "letters from space" as well. While the majority of *Silent Spring* is meticulously researched and always conservative in its use of scientific research, the initial chapter and a small number of other sections of the book move from the scientific to a narrative in which, as Joshua David Bellin notes, "Carson freely admits a lack of conclusive evidence." With those sections, "the speculative nature of her rhetoric . . . distinguishes Cason as a science fiction writer" (Bellin 2009, 146). Where O'Neill's High Frontier will offer colonists a form of total control of "nature," Carson's narrative operates, as Jones and Wills note, "principally through seeing *nature* at risk" (2005, 84; original emphasis). These dystopian speculative sections of *Silent Spring* in which Carson "draws upon narrative traditions from the fable to the jeremiad to the anecdote to the scientific paper" resemble similar sections in Ehrlich's *The Population Bomb* will quickly become a common trope in the Models of Doom genre (Foote 2007, 742). At the same time, sections like "A Fable for Tomorrow" combine, with the rest of *Silent Spring*, to offer a narrative structure that is surprisingly similar to O'Neill's use of scientific plausibility combined with fictional narratives in *The High Frontier*.

Yet, despite the science fictional rhetoric employed by Carson the public, it seemed, was ready for her book. U.S. Supreme Court associate justice William O. Douglas wrote to Carson in early 1962, inquiring after the publication timetable, and would eventually write a glowing review that would be included in the Book-of-the-Month Club edition (Graham 1970, 46). Even before the publication of the book, *The New Yorker* serialized one-third it, and the public was already paying close attention to Carson's argument. In an August 29, 1962, news conference with President John F. Kennedy a reporter referencing the "growing concern among scientists as to the possibility of dangerous long-range side effects from the widespread use of DDT and other pesticides," asked if Kennedy had considered asking the Department of Agriculture or the Public Health Service to examine the issue. Kennedy, obviously familiar with *The New Yorker* articles as the book had yet to be published, replied, "Yes, and I know that they [scientists] already are. I think particularly, of course, since Miss Carson's book, but they are examining the matter " (Kennedy 1962). Under the direction of the president, in May 1963 the Presidential Science Advisory Committee had reported to the Oval Office that Carson was largely correct in her assessment of DDT and other pesticides. The forty-three-page report called for an immediate reduction in the use of pesticides, acknowledging that *Silent Spring* had made the public aware of their toxicity (Presidential Science Advisory Committee 1963).[7]

However, *Silent Spring* did not simply argue that the use and overuse of chemical pesticides on nature might be a harmful activity but it strongly suggested on a deeper level that the very idea of modern technological progress might bring the eventual downfall of humanity. For Carson, DDT was an example of what she called the "Needless Havoc" of human activity. As a result, her main point was not, wholly, about chemical pesticides. That underlying argument was about the behavior of modern humans. For Carson the overuse of chemical pesticides was merely another case of a long list of examples of humanity destroying, needlessly, the natural world. Two early working titles for *Silent Spring* provide a less poetic, but more specific, notice of her theme. Called *Man's War against the Earth* and *The War against Nature*, Carson may have eventually accepted a different title, but the connection to a more general statement about the destruction of nature remained in the text (cf. Oravec 2000, 48–49; Lear 1997, 324). She explained that connection this way:

> As man proceeds towards his announced goal of the conquest of nature, he has written a depressing record of destruction, directed not only against the earth he inhabits but against the life that shares it with him. The history of the recent centuries has its black passages—the slaughter of the buffalo on the western plains, the massacre of the shore-birds by market gunners, the near-extermination of the egrets for their plumage. Now, to these and others like them, we are adding a new chapter and a new kind of havoc—the direct killing of birds, mammals, fishes, and indeed practically every form of wildlife by chemical insecticides indiscriminately sprayed on the land.
>
> (Carson 1999, 87)

For Carson, then, DDT represented merely one more attempt at the subjugation of the natural world—a natural world on which humanity's survival depended. She left no doubt as to her position on the paradigm in which the indiscriminate use of science and technology to remake the planet for the short-term needs of humanity. *Silent Spring* was more than simply a work of science writing, and while its thesis was about pesticides, its overarching theme was far more complex. And, it was also a critique of traditional American frontier writing as well. Where, for example, that American literature scholar Annette Kolodny could find in nineteenth-century women's frontier writing a "voice of *both* adventurer *and* domesticator," Carson worried about the scientific methods for domesticating that frontier (Kolodny 1984, 240; original emphasis). To advance her thesis about this concern Carson suggested exchanging the term pesticide with "biocide" and argued, convincingly to many, that the overuse of such products would soon bring unimaginable destruction of both the natural world and the human environment that depended on it. As historian Mark Hamilton Lytle has noted, with *Silent Spring* Rachel Carson had not simply written a book about the effects of pesticides on the environment but had "quite self-consciously decided to write a book calling into question the paradigm of scientific progress that defined postwar American culture" (2007, 166–167). She ends the book with a rhetorical flourish against that very idea of dominion over nature. And, such dominion remains central to the frontier ideology O'Neill would advance:

> The "control of nature" is a phrase conceived in arrogance, born of the Neanderthal age of biology and philosophy, when it was supposed that nature exists for the convenience of man. The concepts and practices of applied entomology for the most part date from that Stone Age of science. It is our alarming misfortune that so primitive a science has armed itself with the most modern and terrible weapons, and that in turning them against the insects it has also turned them against the earth.
>
> (Carson 1999, 257)

Rachel Carson did this by both employing her skills as a trained biologist and a lyrical writing style to describe a possible future that was not enticing at all. Carson carefully wrote and rewrote in a fashion that M. Jimmie Killingsworth and Jacqueline S. Palmer noted, "invokes the specter of 'evil science' (embodied in popular culture as the mad scientists of comic books and the egghead aliens of science fiction movies)" (1996, 30). Her rhetoric is as much, as Killingsworth and Palmer claim, a "brief experiment in science fiction" as it is about informative science writing (2000, 177). For Killingsworth and Palmer (1996), Carson specifically employed the rhetoric and techniques of apocalyptic science fiction to undermine both the notion of continual human progress and total mastery over the natural environment.

That apocalyptic rhetoric worked for Carson, as literary scholar Lawrence Buell noted, because her style employed cultural knowledge in which, "the apocalyptic frame of reference that high-tech militarism and years of Cold War consciousness [had] implanted in her readers' minds" (Buell 1995, 293). Carson was not just writing a story of the misuse of pesticides; she was writing of a possible future apocalypse. Moreover, in the early 1960s the apocalypse was a narrative of war fought with nuclear bombs. What Carson did was write a story about pesticides that made them, as

Ralph H. Lutts notes, "a kind of pollution that was invisible to the senses" that "could result in cancer, birth defects, and genetic mutations." As a result, the chemical "biocides" of *Silent Spring* were, rhetorically, closely connected to a Cold War apocalyptic imagination, for "pesticides could be understood as another form of fallout" (2000, 19).

Debates about the specific rhetorical style employed by Carson center on whether she "chose to employ the device of mythic storytelling to engage her readers and heighten their concern" or to merely present "a gripping plot and one that is well known to writers and readers of modern science fiction" (Oravec 2000, 46; Killingsworth and Palmer 2000, 178). Regardless of which analysis is correct most agree that, with *Silent Spring*, Rachel Carson presented a narrative in which "scientific humanity is overwhelmed by its own power to control nature, and the Fall of humankind remains a possibility even as science seeks human perfectibility" (Killingsworth and Palmer 2000, 178). This is, of course, precisely the opposite theme of O'Neill's High Frontier, where human perfectibility is made possible, exclusively, although a combination of science and the control of the natural world. And it is this narrative contrast that makes O'Neill's work striking in its utopian vision.

At the same time, both O'Neill's *The High Frontier* and Carson's *Silent Spring* share a common enough starting point. What Frank Graham, Jr., noted about Carson's work actually applies equally—but to different purposes—in O'Neill's. Graham claimed that *Silent Spring* set out to use "science as its base" but "nonetheless transcend[s] those limited confines of the average scientist's mind which had pulled the world into its current morass" (1970, 21). However, *Silent Spring* goes far beyond the use of chemical pesticides and advances a thesis that directly contradicts the one O'Neill will eventually put forth, that "the human attempt to control nature, to improve on nature, leads finally to the death of nature" (Killingsworth and Palmer 2000, 178). For Carson humanity found itself confronted with an ominous and possibly apocalyptic choice. The early 1960s image of science as a benign source of human advancement was increasingly fraught with peril. It was a road, Carson wrote, "we have long been travelling [that] is deceptively easy, a smooth superhighway on which we progress with great speed, but at its end lies disaster." On that road lay the overuse of chemical pesticides, and more generally the paradigm of continuing scientific progress. The other direction, the "one 'less travelled by'" Carson notes, citing the well-known Robert Frost poem, "offers our last, our only chance to reach a destination that assures the preservation of our earth" (1999, 240).

In a way, then, Gerard K. O'Neill's reaction to works such as *Silent Spring* is to misread them. While his *The High Frontier* presents a narrative featuring many of the same rhetorical strategies derived from science fiction as *Silent Spring*, his conclusions ignore Carson's main point: that science and technology may be the original source of the problem. For O'Neill, however, science and technology are more than a possible solution; they are the *only* solution. Later in his career he would put this paradigm quite bluntly, arguing in the preface to his 1983 pro-science apologetic *The Technology Edge*, his belief that humanity

> will continue to make progress. The degree of control they exercise over the Earth will increase. They have the responsibility to be the saviors of all other creatures . . . and because human beings pursue ideals, eventually those ideals will come true.
>
> (O'Neill 1983, 8–9; the ellipses are in the original)

In the orbiting colonies of O'Neill's High Frontier science and technology can be used not only to control nature but also to re-create a new form of nature—and for O'Neill this is, indeed, an "ideal." Where Carson is most concerned with the effects of attempts to control the natural world, O'Neill sees his orbiting frontier as a way to control it completely. Where Carson specifies the negative effects of chemical pesticides like DDT, O'Neill notes that in orbit there will be no need for such earthly concerns.

It is with his discussion of agriculture that O'Neill makes this point most clear. "For the first time," he writes, "we will be able to have agriculture of high yield without pesticides, insecticides, or crop losses due to raiding birds and animals" (O'Neill 1976b, 51). In the orbiting frontier of O'Neill's colonies the specific concerns of Rachel Carson on chemical pesticides will be eliminated simply because there will be no pests. O'Neill presents a narrative not of a "Silent Spring" but of a spring dominated only by those birds chosen by humans for their beautiful songs. His version of nature is one that can—and should—be totally controlled. O'Neill writes that "[o]ur Earth is rich in plants and animals, but as industry and human population crowd environments it is not as rich as it once was." Nature, however, can be easily recreated in space, and for the simple reason that humans enjoy it. O'Neill notes that "city children become starved for the sight of a tree, and in desert areas the palms of oases have an importance no dweller in a lush climate can imagine" (1976b, 48–49). This need for the natural necessitates its total control in the High Frontier. Humans will take nature with us to the next frontier because it is useful to us and because we simply want it:

> For our psychological well-being, as well as for the cycling of oxygen we breathe, we should have grass, trees, and flowers. Many animal species are a pleasure to us, and if we move into space both we and they will benefit by our taking them along . . . Along with the domestic animals, we will certainly want to bring squirrels, deer, otter, and many others; birds, and some types of harmless insects for them to eat. In space, though, we have an option that does not exist to us on Earth: to take along those species which we want and which form parts of a complete ecological chain, but to leave behind some parasitic types: how delightful would be a summertime world of forests without mosquitoes!
>
> (O'Neill 1976b, 49)

Where Rachel Carson worried, as American literature scholar Michael A. Bryson put it, that "the Western impulse since World War II to control nature, to selectively destroy certain organisms (primarily insects) . . . is based on profits and convenience rather than sound" science, Gerard K. O'Neill firmly remains in the modernist camp of science as force for progress—and human convenience (Bryson 2002, 165). In O'Neill's frontier utopia there will be birds and "there must be insects, perhaps butterflies, for the birds to eat, but there need not be mosquitoes—or cockroaches, or rats" (1976b, 221). The colonists of O'Neill's colonies will exert total control over the natural environment they have re-created from desired forms from Earth. Everything entering the specific orbital environments built for agricultural use will be pass "through a high temperature solar furnace" rendering them sterile. "In that way that can be kept free of pests, even is any should accidentally be introduced to the living habitat" (O'Neill 1976b, 76). However, lest the reader get the wrong impression about the level of control of the environment in his High Frontier, O'Neill offers the perfect form of pest control as well as a decisive statement about his opinion of the relationship between humanity and nature:

> In the very worst case—the introduction or evolution of an agricultural disease in one of the areas—the sterilization process that will be part of the recycling will ensure that the disease will not spread. As soon as such a disease is found, there will be a simple and preferable alternative to the Earthbound necessity of sprays and poisons: it will only be necessary to drain off the water of the contaminated cylinder through a solar steam boiler to a sterile tank, and to open the shades so that the cylinder heats up to a temperature in which no living organism can survive. After a few days or weeks of that treatment the water can be re-introduced, the appropriate soil bacteria can be replaced, and a new planting cycle can begin.
>
> (1976b, 76)

Yet, if O'Neill offered the ability for the total control of "nature" in his orbiting habitats, sterilizing anything, which presented a potential danger, the end result was an ability to ignore the "nature" of the human beings his islands in space were designed for. Another Model of Doom author, Paul R. Ehrlich, would suggest that an inability to control "human nature"—that is, reproduction—was a threat equal to, if not surpassing, Carson's. Thankfully, O'Neill's utopia could solve that problem, too.

DEFUSING *THE POPULATION BOMB*

With his Island Three Gerard K. O'Neill envisioned a future space colony with as many as several million inhabitants. But for a planet with a population in the early 1970s just approaching four thousand million the idea that such colonization efforts could stem the population explosion might seem fanciful. Yet, O'Neill suggests precisely that. His utopian rhetoric counters the dystopian vision of another Model of Doom proponent, Paul R. Ehrlich. Ehrlich's *The Population Bomb*, first published in 1968, became an almost instant international best seller (Ehrlich 1968).[8] A biologist at Stanford University, Ehrlich specialized in the study of butterfly populations and genetics and turned that specialty toward the human population and its continual growth. Written for a general audience, *The Population Bomb* made no effort to suggest the problems Ehrlich outlined had any kind of solution. A year before the publication of his best-selling work, Ehrlich made his case in an article in *New Scientist*, claiming that "in the 1970s and 1980s hundreds of millions of people will starve to death in spite of any crash programs embarked upon now" (1967, 653). *The Population Bomb* would be a detailed description of precisely how those "hundreds of millions" of deaths would occur.

In a true neo-Malthusian fashion, Ehrlich declared that the Earth's human population was about to face massive starvation and that nothing could forestall it. "More than half the world is in misery now," he noted, and "that alone should be enough to galvanize us into action, regardless of the exact dimension of the future disaster now staring *Homo sapiens* in the face" (Ehrlich 1968, 189). The "future disaster" was, however, inevitable. Ehrlich offered no solution, only plans to reduce its impact. As with other Models of Doom writers, for Ehrlich, his central conundrum, was—although unstated—a lack of resources and a lack of frontiers. The human population was expanding too quickly, and there were no new frontiers for those far too many humans to exploit.

Like Rachel Carson, Ehrlich used the dystopian rhetorical strategy of presenting an apocalyptic future. Indeed, later Paul R. Ehrlich along with his frequent collaborator and wife, Anne H. Ehrlich, would eventually specify *Silent Spring* as an important precursor to *The Population Bomb* (Ehrlich and Ehrlich 2009, 62). Yet, unlike Carson's more general "Fable for Tomorrow," Ehrlich was far more specific. Presenting three scenarios that "describe the kinds of disasters that *will* occur as mankind slips into the famine decades" Ehrlich takes his place as one of the most apocalyptic of the doomsday authors (1968, 39; original emphasis). His style is in line with Carson's, as an "experiment in science fiction," and his scenarios are written in the present tense, as though Ehrlich is merely reporting on the apocalyptic events he foretells. Writing in 1968 Ehrlich begins his first scenario in his reader's very near future, 1972, with "catastrophic floods" in China, war in Thailand, and "food riots" in China, India, and Brazil (39–40). At the end of Ehrlich's "Scenario I" an American president has called for a nuclear attack against "Chinese air bases, supply complexes, and staging areas in North Vietnam" and "with the connivance of the Russians a pre-emptive nuclear strike is also launched against China's nuclear facilities" (Ehrlich

1968, 40). By the end of Scenario I "more than 100 million Americans" are dead in a nuclear war fought, essentially, over limited world food resources.

But Scenario I is not nearly the worst possible future Ehrlich envisions. In Scenario II, set in 1979, Ehrlich begins with the last noncommunist government in Latin America, Mexico, being "replaced by a Chinese-supported military junta" (Ehrlich 1968, 40). As food riots in the 1970s sweep across Africa, Asia, and South America, Ehrlich notes that "food riots have often become anti-American riots, as [American] enemies claimed it was withholding food from the starving" (1968, 40). A strange sort of apocalyptic luck is the only thing that prevented nuclear war over food in Ehrlich's 1970s as "the outbreak of a particularly virulent strain of bubonic plague killing 65 per cent of the starving Egyptian population had averted a direct Soviet-American clash in the Mediterranean" (Ehrlich 1968, 41). By the end of Scenario II "killer smog" has caused the deaths of ninety thousand Los Angelinos, the polar ice caps are melting, the American president is considering mandatory sterilization for all Americans with an IQ under 90 and, soon, nuclear war will be inevitable. "The most intelligent creatures" that will survive these events are, of course, cockroaches (Ehrlich 1968, 42–43).

Ehrlich is somewhat more optimistic in his Scenario III. American "expeditionary forces" are withdrawn from Thailand and Vietnam by 1974, the pope announces the Catholic Church's acceptance of birth control and abortion, and as famine sweeps the Arab world, they seek peace with Israel and bring stability to the Middle East. Japan and Australia form a pro-American military alliance bringing similar stability to Southeast Asia and the Pacific. As a result of this stability, the United Nations, with support of both the United States and the Soviet Union, as well as, "Japan, Australia, and the Common Market [European] countries," initiates a plan that will "involve simultaneous population control, agricultural development, and limited industrialization, to be carried out jointly in selected sections of Asia, Africa, and South America" (Ehrlich 1968, 44).

For Ehrlich, Scenario III is his attempt at optimism. He notes that it "has considerably more appeal than the others, even though it presumes the death by starvation of perhaps as many as 500 million people, one-seventh of the world's population" (Ehrlich 1968, 44). He then challenges his readers to "create a more optimistic" future than his Scenario III, pointing out that he will not "accept one that starts, 'In early 1972 the first monster space ships from a planet of the star Alpha Centauri arrive bearing CARE packages'" (Ehrlich 1968, 44). Indeed, in creating his three scenarios Ehrlich is specifically extrapolating an apocalyptic future for the intended purpose of showing his readers how dim their prospects are. There are, simply, too many people and no new frontier in which to place them for Ehrlich. And, in presenting his three scenarios he presents arguments strikingly similar to those of Rachel Carson and The Club of Rome. For the human species to survive, Ehrlich says that "we've got to change from a growth-oriented, exploitative system to one focused on stability and conservation" (1968,

106). Although he does not put it in the terminology of traditional frontier mythology, a change from a "growth-oriented, exploitative system" is, in reality, a move away from what others would call a "frontier ideology."

Tracing the general trends in both world population growth and food production, and projecting them into the near future, Ehrlich predicted hundreds of millions and, perhaps, billions of deaths due to malnutrition and starvation in the very near future. For Ehrlich, the growth in the human population during the twentieth century was the major cause for all of the world's problems. Population growth was to blame for everything from wars over resources to environmental degradation. As nations used increased levels of pesticides throughout the world in attempts to create more farmland, it was the always-growing human population that was responsible for the majority of ills as outlined by Rachel Carson in *Silent Spring*. Attempts to force farmland to produce more than it could, and implement new farming technologies and strategies were only pushing off the inevitable tragedy of mass starvation as those new farming methods were eventually doomed to fail. But, while they worked they would only provide an impedance for continued population growth, making the inevitable tragedy all the worse.

It is impossible to understate the overheated rhetoric of Paul Ehrlich's *The Population Bomb*: it was a book designed to proclaim that the "end" was very near. Offering no real solutions; it only predicted death and degradation through overpopulation. While it offered various scenarios of, precisely, how the massive human "die off" would occur, for Ehrlich it was an inevitable eventuality. At most, through massive governmental intervention, it might be possible to reduce the number of deaths by starvation, but such deaths would remain massive in numbers. His prescription for the 1970s and after was apocalyptic:

> The rich may continue to get richer, but the more numerous poor are going to get poorer. Of these poor, a *minimum* of ten million people, most of them children, will starve to death during each year of the 1970s. But this is a mere handful compared to the numbers that will be starving before the end of this century. And it is now too late to take action to save many of those people.
>
> (Ehrlich 1968, 3; original emphasis)

Looking toward demographic trends, Ehrlich saw the same thing Thomas Malthus had envisioned almost two centuries before, namely that the human population was continuing to grow exponentially while food production would grow, at best, arithmetically. As a result, he predicted an inevitable moment when the human population would be reduced through starvation. Importantly, Malthus failed to account for technological change and expanding frontiers in his formulae. O'Neill will argue, with his High Frontier, that Ehrlich made the same mistake.

In an effort to prove the inevitability of massive starvation, Ehrlich turned to the work of British physicist J.H. Fremlin. As a thought experiment for the popular science magazine *New Scientist*, Fremlin worked out the needs of a continuing expansion of the human population at early 1960s rates on the Earth and extrapolated that population growth into the wider solar system (Fremlin 1964).[9] Fremlin dutifully provided calculations of the land area for all the inner planets of the solar system, as well as the moons of the outer planets Jupiter and Saturn. Ehrlich, citing Fremlin, notes that at the current rate of population growth (in 1968), "it would take only about 50 years to populate Venus, Mercury, Mars, the moon, and the moons of Jupiter and Saturn to the same population density as Earth" (1968, 5–6).

Paul Ehrlich's use of Fremlin's logic was simply a rhetorical attempt at hyperbole. Ehrlich hoped to convince his readers that expansion into space was not a solution to the population bomb he was describing. After detailing and expanding upon Fremlin's *New Scientist* essay, Ehrlich declared,

> Enough of fantasy. Hopefully, you are convinced that the population will have to stop growing sooner or later and that the extremely remote possibility of expanding into outer space offers no escape from the laws of population growth.
>
> (1968, 6)

Gerard K. O'Neill, as before, had an answer for Paul Ehrlich, too. His "Colonization" essay stated, at the outset, that "[t]he ultimate size limit for the human race on the newly available frontier is at least 20,000 times its present value" (1974, 37). O'Neill's High Frontier was more than the classic frontier as "social safety-valve" envisioned by Frederick Jackson Turner. It was a complete population safety valve, allowing for near near-term expansion into orbit and, because of what that expansion offered Earth, a way to curb population growth without Ehrlich's massive "die-off." While Ehrlich would not accept a scenario of aliens with CARE packages, the scenario O'Neill offers in the High Frontier merely removes the aliens, and replaces them with industry and new territory in Earth orbit. Building materials from the moon, and later the asteroids, allow for an ongoing process of building new islands in space. Carefully presenting his arithmetic, O'Neill suggests that his High Frontier will allow both continual growth of the human population and a place to house them all with ever-increasing standards of living. For O'Neill, the opening of the High Frontier would, "by about 2050," defuse the population bomb:

> emigration to the colonies course reverse the rise in Earth's population, and . . . the acceleration of the solution could be dramatically fast: Within less than 30 years, Earth's population could be reduced from a peak of 16.5 billion people to whatever stable value is desired. I have

> suggested 1.2 billion as a possible optimum; it corresponds to the year 1910 in Earth history.
>
> (1974, 37)

In part Ehrlich's population bomb is defused by providing room to expand. More centrally to O'Neill's argument, however, is the cheap energy provided by his solar power satellites, the movement of industry from the surface of the Earth to orbit, and from them a continual increase in the standard of living for all humans, everywhere. The surest way to reduce the rate of population growth, O'Neill notes, is through a rising standard of living. As a result, O'Neill's argument for a reduction in population growth hinges on "an increase [of] the wealth of the underdeveloped countries not just by a few percent per year but massively, by factors of ten or a hundred" (1976b, 30). Through such a plan, a primary effect of the High Frontier, "would only require thirty years from the completion-date of the first community before new lands would be increasing more than fast enough to cope with such [population] demands" (O'Neill 1976b, 261).

LIGHTING OUT FOR THE ORBITAL UTOPIAN TERRITORY

In a final collation of the work of people such as Carson and Ehrlich, the Club of Rome would publish a complete testament to the need to end frontier ideology in 1972 with *The Limits to Growth*. Using what was, for the time, a complex computer model the *Limits* authors forcefully proclaimed an end to frontiers and the need for growth inspired by them. The authors' logic was simple:

> If the present growth trends in world population, industrialization, pollution, food production, and resource depletion continue unchanged, the limits to growth on this planet will be reached sometime with the next one hundred years.
>
> (Club of Rome 1972, 23)

Thus, The Club of Rome called for, simply, the opposite of frontier ideology. It foresaw a "period of great transition," away from any expanding frontiers and toward "the transition from growth to global equilibrium" (Club of Rome 1972, 24). O'Neill's solution was a different kind of transition altogether—one of moving the frontier off the surface of the planet where "the limits to growth on this planet" were not a concern. As Howard P. Segal put it, O'Neill's vision was both "a hardnosed technical work" like that of the *Limits* writers, but at the same time was "a passionately romantic one" (1985, 152). Nor would there be a need to worry about problems with environmental toxins, population expansion, or any other of a veritable plethora of prophesied coming disasters for the human species by the Models of Doom writers.

It is possible to show in detail how any of the Models of Doom authors' scenarios are countered by O'Neill's High Frontier. O'Neill's utopia makes the use of science that Carson warned about a far more benign force. With continually expanding colonies, first in orbit and then throughout the solar system, there is no need to worry about the human population bomb. Additionally, without competition for the limited land area and limited recourses the possibility of nuclear war is greatly reduced. Without ever declining sources of energy thanks to solar power satellites, the human population can easily and continually increase its standards of living and, because of this, the freedom of choice for each individual. O'Neill's solution to all of the Earth's perceived problems become solved once humans leave their home planet and, like European explorers and colonists centuries before, begin creating wealth on the High Frontier.

In 1985 a number of the Models of Doom scholars and authors met at Texas A&M University for a conference on their continuing work. Recalling the old myth of the Greeks in which a woman was condemned to have the gift of prophecy but be continually disbelieved whenever she foretold the future, they called it "The Cassandra Conference" (cf. Ehrlich and Holdren 1988b). One of the original authors of *The Limits to Growth* report, Donella H. Meadows, offered a succinct history of what the conference attendees saw as the apocalyptic future awaiting humanity. Citing Paul Ehrlich (who would edit the conference papers), Rachel Carson, and a significant list of other writers as people aiming "literary grenades tossed from the sidelines at the assumptions of . . . the establishment," Meadows provides an excellent list of authors in the Models of Doom genre and, in doing so, also offered no easy solutions (1988, 257). She also summarizes the main lesson of that genre, claiming, "there are lessons to be learned from these books: lessons about the larger society and its ways of handling unorthodox ideas" (Meadows 1988, 258). For Meadows the "unorthodox ideas" have not changed since Carson published *Silent Spring*, Ehrlich published *The Population Bomb*, and the Club of Rome published *The Limits to Growth*. Without radical changes the human species is destined to a very apocalyptic future, with starvation, environmental degradation, war, and possibly extinction. It is necessary, Meadows told her fellow Cassandras in 1985, to offer plans to force the "stabilization of the human population . . . the stabilization of the annual flow of energy and materials through the economy . . . [and] the just distribution of both material and nonmaterial goods so that every member of the human race is included, supported, acknowledged, and enabled to contribute fully to [this] joint enterprise" (1988, 268).

For Meadows—and, indeed, all the attendees of the conference, and all the Models of Doom writers—gone were the days when simply finding new frontiers offered new opportunities, new wealth, and expanded freedom. While "the earth is a fruitful, beautiful, and resilient planet," and humans are "a strong and ingenious species," the dangers presented by Meadows and her colleagues, they believed, simply had to be overcome. "If the global society just decided to do it," Meadows acknowledged, "it would be simple to create a secure, equitable, sustainable, and sufficient life for everyone." But to accomplish that it was necessary to also acknowledge that the changes, though simple, required the admission that it was no longer possible to "maintain continuous physical growth *on a finite planet*" (Meadows 1988, 269; emphasis added). What was needed, simply, was an economy, a model of living, and a mind-set that could best be described as "steady-state," or, to put it more bluntly, that frontiers—and frontier growth—were now impossible for the human species.

Not surprisingly for Gerard K. O'Neill the mistake made by Models of Doom scholars and authors was that final assumption—that growth was impossible, the planet Earth finite, and frontiers were no longer part of the human equation. Marrying the science and technology of NASA to the frontier myths of *Star Trek* and, indeed, the frontier myths of the American nation, O'Neill's solution was exclusively about frontiers. Where the Cassandra Conference attendees firmly believed that it was necessary "to make choices that can improve [the] prospects for a prosperous and secure world," O'Neill believed the choice was about leaving that world behind and beginning anew to create his utopia (Ehrlich and Holdren 1988a, 317–318). So, really, O'Neill's solutions were about the myth of American frontiers. He might have repeatedly claimed that his vision was not utopian, but it also offered solutions to every problem he believed beset humanity. Just four years before the Cassandra Conference O'Neill had offered his own prophecy of the future, and where the prophecies of the Cassandras were prophecies of doom, his was far more optimistic:

> America is a frontier society lacking a frontier, and I suspect that when space is opened for settlement . . . individual Americans will move into space with a vigor and drive that will astonish us . . . By 2081 there may be more Americans in space colonies than there are in the United States."
>
> (O'Neill 1981, 71)

While O'Neill firmly thought that "thousands of people now alive may choose within the next two decades to live and work on the new frontier of space," much of that optimism was built to be conveyed into orbit by the space shuttle (1976b, 3). However, when the shuttle failed to live up to expectations O'Neill's High Frontier became impossible. The "100–200 flights" of the shuttle in the 1980s that Eric Drexler wrote of in 1977 became a total of 135 flights during the entire operational life of the shuttle through 2011. After the loss of the *Challenger* in 1986 it became clear, as T. A. Heppenheimer put it, that

> [t]he shuttle could not be all things to all people. In no way was it capable of the rapid and routine operational uses that its builders had

promised. It had carried hope of frequent and low-cost space operations, but now this hope was dead.

(2002, 413)

Without the shuttle delivering as promised, the "second-generation" shuttles, and "shuttle-derived vehicles" that O'Neill could use in his planning for the 1990s would never materialize. Thus, there could be no exodus of the population into the High Frontier, no solar powered satellites beaming energy back to Earth, and as a result, no frontier utopia at all.

But to doubt that O'Neill's High Frontier is anything other than a restatement of an ongoing belief in the purpose and the promise of frontier myths for Americans is to truly doubt the continuing strength of those frontier myths, themselves. Terrence Ball described O'Neill's High Frontier in a way that captures this belief. Ball writes,

> The geographic and physical frontier of the American West may well be closed forever. But for Americans the *idea* of the frontier is not, and never will be, dead or *passé*. And this is not an idea that exists exclusively in the historical past—an abomination for Adam—but in a present perpetually alive with possibilities for self-assertion and exploration . . . [The frontier] is in its very nature infinitely open, presenting an unlimited horizon of limitless possibility, of renewal, regeneration, and rebirth. Americans need never confront the closure and finitude of the 'steady state' society and an economy of scarcity. That being the case, they are forever able to strike out, in Huck Finn fashion, for the territories, thereby thumbing their noses at Aunt Sally—and Hegel—and anyone else who wants to 'sivilize' us by restricting our freedom.
>
> (1995, 287–288; original emphasis)

In a fashion, however, what Ball writes of O'Neill could also apply to all the frontiers within this book. Be they the lost frontiers of Haldeman and the Vietnam War, the hoped-for frontiers of NASA and Apollo, the fictional final frontier of Captain Kirk and the *Enterprise*, or the utopian High Frontier of Gerard K. O'Neill, the continuing significance of those frontiers persist. That O'Neill—or any of these other experiments in mythmaking—was forced to concede to technological, practical, and empirical constraints does not indicate a need to also acquiesce to limits on the mythical boundaries of such frontiers.

NOTES

1 Drexler (1977, 104–105). Drexler, later publishing as K. Eric Drexler, would finish his PhD at the Massachusetts Institute of Technology in 1991 on nanotechnology.

2 Pages 139 through 146 discuss the idea of a "mass driver"; pages 54and 55 offer specific details on "solar power satellites."

3 Dyson provides a complete and accurate overview of O'Neill's professional and personal life and biographical information is drawn largely from this as well as O'Neill's own autobiographical statements, unless otherwise specified.
4 The spelling of Tsiolkovsky is an acceptable Anglicization of the period; the characterization of Clarke's novel as being for children is incorrect.
5 Said is referring to his idea of "Orientalism," but the statement can be taken metaphorically—as much of Said's work has been, previously.
6 A surprising number of the early English explorers and colonists cited in the chapter on NASA could be cited here as well. See, especially, the section "Bradford and the New World as Metonymic Signifier for Apollo."
7 The significance of this report is such that it remains on permanent public display in the atrium at the John F. Kennedy Presidential Library.
8 Later Ehrlich would note that authorship should have included his wife, Anne H. Ehrlich. The book will be referenced as published in 1968.
9 Fremlin was a nuclear physicist, and his primary point was that, eventually, the human population would be so large that it would reach a "heat limit" for the Earth in which the heat generated by the human population would exceed the ability of the planet to stabilize. Fremlin's essay also played an important inspirational role in the novel *Make Room! Make Room!* (Harrison 1966), the basis for the 1973 Charlton Heston film *Soylent Green* (Fleischer 1973).

WORKS CITED

Bacon, Francis. 1915. *New Atlantis*. Oxford: Clarendon Press.

Ball, Terence. 1995. *Reappraising Political Theory: Revisionist Studies in the History of Political Thought*. Oxford: Clarendon Press.

Bell, Daniel. 1973. *The Coming of Post-Industrial Society: A Venture in Social Forecasting*. New York: Basic Books.

Bell, Daniel, and Stephen R. Graubard, eds. 1968. *Toward the Year 2000: Work in Progress*. Boston: Houghton Mifflin.

Bellin, Joshua David. 2009. "Us or Them!: Silent Spring and The 'Big Bug' Films of the 1950s." *Extrapolation* 50 (1): 145–70.

Billington, Ray Allen. 1973. *Frederick Jackson Turner: Historian, Scholar, Teacher*. New York: Oxford University Press.

Brand, Stewart. 1977. "Is The Surface of a Planet Really the Right Place for an Expanding Technological Civilization? (Interview with Gerard O'Neill)." In *Space Colonies: A Co-Evolution Book*, edited by Stewart Brand, 22–30. Sausalito, CA: Whole Earth Catalog.

Bryson, Michael A. 2002. *Visions of the Land: Science, Literature, and the American Environment from the Era of Exploration to the Age of Ecology*. Charlottesville: University Press of Virginia.

Buell, Lawrence. 1995. *The Environmental Imagination: Thoreau, Nature Writing, and the Formation of American Culture*. Cambridge, MA: Belknap Press of Harvard University Press.

Carson, Rachel. 1999. *Silent Spring*. New York: Penguin Books.

Club of Rome. 1972. *The Limits to Growth: A Report for the Club of Rome's Project on the Predicament of Mankind*. Edited by Donella H. Meadows, Dennis L. Meadows, Jørgen Randers, and William W. Behrens III. New York: Universe Books.

Desrochers, Pierre, and Christine Hoffbauer. 2009. "The Post War Intellectual Roots of the Population Bomb. Fairfield Osborn's 'Our Plundered Planet' and William

Vogt's 'Road to Survival' in Retrospect." *The Electronic Journal of Sustainable Development* 1 (3): 37–61.

Discover Magazine Editors. 2006. "25 Greatest Science Books of All Time." *Discover Magazine*, December. http://discovermagazine.com/2006/dec/25-greatest-science-books/article_view?b_start:int=1&page=2.

Drexler, Eric. 1977. "The Space Colonies Idea, 1969–1977." In *Space Colonies: A CoEvolution Book*, edited by Stewart Brand, 104–8. Sausalito, CA: Whole Earth Catalog.

Dyson, Freeman J. 1993. "Gerard Kitchen O'Neill (Obituary)." *Physics Today* 46 (2): 97–98.

Ehrlich, Paul R. 1967. "Paying the Piper." *New Scientist* 36 (December): 652–55.

———. 1968. *The Population Bomb*. New York: Ballantine Books.

Ehrlich, Paul R., and Anne H. Ehrlich. 2009. "The Population Bomb Revisited." *The Electronic Journal of Sustainable Development* 1 (3): 62–71.

Ehrlich, Paul R., and John P. Holdren. 1988a. "Editor's Epilogue." In *The Casandra Conference: Resources and the Human Predicament*, edited by Paul R. Ehrlich and John P. Holdren, 317–18. College Station: Texas A&M University Press.

———., eds. 1988b. *The Cassandra Conference: Resources and the Human Predicament*. 1st ed. College Station: Texas A&M University Press.

Fleischer, Richard. 1973. *Soylent Green*. Hollywood: Metro-Goldwyn-Meyer.

Foote, Bonnie. 2007. "The Narrative Interactions of Silent Spring: Bridging Literary Criticism and Ecocriticism." *New Literary History* 38 (4): 739–54.

Freeman, Christopher. 1973. "Part I: The World Models and Their Sub-Systems." In *Models of Doom: A Critique of the Limits to Growth*, edited by H S D Cole, Christopher Freeman, Maria Jahoda, and K L R Pavitt, 5–13. London: Chatto & Windus.

Fremlin, J. H. 1964. "How Many People Can the World Support?" *New Scientist* 24 (3): 285–87.

Frye, Northrop. 1973. "Varieties of Literary Utopias." In *Utopias and Utopian Thought*, edited by Frank E. Manuel, 25–49. London: Condor.

George, Susan A. 2000. "Space for Resistance: The Disruption of the American Frontier Myth in 1950s Science Fiction Films." In *Space and Beyond: The Frontier Theme in Science Fiction*, edited by Gary Westfahl, 77–83. Westport, CT: Greenwood Press.

Glaser, Peter E. 1968. "Power from the Sun: Its Future." *Science* 162 (3856): 857–61.

———. 1976. "The Status of the Satellite Solar Power Station (AAS 75-007)." In *The Proceedings of the Thirteenth Goddard Memorial Symposium on Future Space Activities of the American Astronautical Society*, edited by Carl H. Tross, 81–102. Tarzana, CA: AAS Publications Office.

Graham, Frank. 1970. *Since Silent Spring*. Boston: Houghton-Mifflin.

Greenblatt, Stephen. 1991. *Marvelous Possessions: The Wonder of the New World*. Oxford: Clarendon Press.

Harrison, Harry. 1966. *Make Room! Make Room!* New York: Doubleday.

Heinlein, Robert A. 1966. *The Moon Is a Harsh Mistress*. New York: G.P. Putnam's Sons.

———. 1977. "Science Fiction: Its Nature, Faults and Virtues." In *Turning Points: Essays on the Art of Science Fiction*, edited by Damon Knight, 3–28. New York: Harper & Row.

Heppenheimer, T. A. 1977. *Colonies in Space*. Harrisburg, PA: Stackpole Books.

———. 1999. *The Space Shuttle Decision: NASA's Search for a Reusable Space Vehicle*. NASA SP 4221. Washington, DC: National Aeronautics and Space Administration, NASA History Office, Office of Policy and Plans.

———. 2002. *Development of the Space Shuttle: 1972–1981*. NASA SP 4221. Washington, DC: National Aeronautics and Space Administration, NASA History Office, Office of Policy and Plans.

Hutton, Sarah. 2002. "Persuasions to Science: Baconian Rhetoric and the New Atlantis." In *Francis Bacon's New Atlantis: New Interdisciplinary Essays*, edited by Bronwen Price, 48–59. Manchester, UK: Manchester University Press.

Johnson, Richard D., and Charles Holbrow, eds. 1977. "Space Settlements: A Design Study (NASA SP-413)." NASA.

Jones, Karen R., and John Wills. 2005. *The Invention of the Park: Recreational Landscapes from the Garden of Eden to Disney's Magic Kingdom*. Cambridge, UK and Malden, MA: Polity.

Kahn, Herman, and Anthony J. Wiener. 1967. *The Year 2000: A Framework for Speculation on the Next Thirty-Three Years*. New York: Macmillan.

Kennedy, John F. 1962. "News Conference (42)." United States Department of State.

Kerridge, Richard. 1999. "Ecologies of Desire: Travel Writing and Nature Writing as Travelogue." In *Travel Writing & Empire: Postcolonial Theory in Transit*, edited by Steve Clark, 164–82. London: Zed Books.

Kilgore, De Witt Douglas. 2003. *Astrofuturism: Science, Race, and Visions of Utopia in Space*. Philadelphia: University of Pennsylvania Press.

Killingsworth, M. Jimmie, and Jacqueline S. Palmer. 1996. "Millennial Ecology: The Apocalyptic Narrative from Silent Spring to Global Warming." In *Green Culture: Environmental Rhetoric in Contemporary America*, edited by Carl G. Herndl and Stuart C. Brown, 21–45. Madison: University of Wisconsin Press.

———. 2000. "Silent Spring and Science Fiction: An Essay in the History and Rhetoric of Narrative." In *And No Birds Sing: Rhetorical Analyses of Rachel Carson's Silent Spring*, edited by Craig Waddell, 174–97. Carbondale: Southern Illinois University Press.

Klein, Richard K. 1977. "Maybe We Are Alone—Physicist Gerry O'Neill Says That's a Reason for Sending People into the Safety of Space." *People Magazine*. http://www.people.com/people/archive/article/0,,20069754,00.html.

Kolodny, Annette. 1984. *The Land Before Her: Fantasy and Experience of the American Frontiers, 1630–1860*. Chapel Hill: University of North Carolina Press.

Kubrick, Stanley. 1964. *Dr Strangelove, Or: How I Learned to Stop Worrying and Love the Bomb*. Burbank, CA: Columbia Pictures.

Lear, Linda J. 1997. *Rachel Carson: Witness for Nature*. 1st ed. New York: Henry Holt.

———. 2000. "Searching for Rachel Carson." In *And No Birds Sing: Rhetorical Analyses of Rachel Carson's Silent Spring*, edited by Craig Waddell, 205–18. Carbondale: Southern Illinois University Press.

Levitas, Ruth. 1990. *The Concept of Utopia*. 1st ed. Utopianism and Communitarianism. Syracuse, NY: Syracuse University Press.

Lockitch, Keith. 2007. "Rachel Carson's Genocide: Carson's Environmental Ideology Demands Opposition to DDT despite the Millions of Malaria Deaths Its Use Could Prevent." *Capitalism Magazine*, May 23. http://www.capitalismmagazine.com/science/environment/pollution/4965-rachel-carson-s-genocide.html.

Lutts, Ralph H. 2000. "Chemical Fallout: Silent Sprint, Radioactive Fallout, and the Environmental Movement." In *And No Birds Sing: Rhetorical Analyses of Rachel Carson's Silent Spring*, edited by Craig Waddell, 17–41. Carbondale: Southern Illinois University Press.

Lytle, Mark H. 2007. *The Gentle Subversive: Rachel Carson, Silent Spring, and the Rise of the Environmental Movement*. New Narratives in American History. New York: Oxford University Press.

Martin, Wendy. 1994. "Introduction: Mapping American Life." In *Colonial American Travel Narratives*, vii–xviii. New York: Penguin Books.

McCay, Mary A. 1993. *Rachel Carson*. New York: Twayne.

McCurdy, Howard E. 1997. *Space and the American Imagination*. Washington, DC: Smithsonian Institution Press.

Meadows, Donella H. 1988. "The Limits to Growth Revisited." In *The Casandra Conference: Resources and the Human Predicament*, edited by Paul R. Ehrlich and John P. Holdren, 257–70. College Station: Texas A&M University Press.

Michaud, Michael A. G. 1986. *Reaching for the High Frontier: The American Pro-Space Movement, 1972–84*. New York: Praeger.

More, Thomas. 2008. *Utopia*. Rockville, MD: Arc Manor.

Moylan, Tom. 1986. *Demand the Impossible: Science Fiction and the Utopian Imagination*. New York: Methuen.

———. 2000. *Scraps of the Untainted Sky: Science Fiction, Utopia, Dystopia*. Boulder: Westview Press.

———. 2003. "State, Agency, and Dystopia." In *Dark Horizons: Science Fiction and the Dystopian Imagination*, edited by Raffaella Baccolini and Tom Moylan, 135–54. New York: Routledge.

Murray, Iain. 2007. "Silent Alarmism: A Centennial We Could Do Without." *National Review Online*, May 31. http://cei.org/op-eds-and-articles/silent-alarmism-centennial-we-could-do-without.

National Review Editors. 1999. "The 100 Best Non-Fiction Books of the Century." *The National Review*, May 3. http://old.nationalreview.com/100best/100_books.html.

O'Neill, Gerard K. 1956. "Storage-Ring Synchrotron: Device for High-Energy Physics Research." *Physical Review* 102 (5): 1418–19.

———. 1974. "The Colonization of Space." *Physics Today* 27 (9): 32–40.

———. 1976a. "Engineering a Space Manufacturing Center." *Astronautics and Aeronautics* 14: 20–28, 36.

———. 1976b. *The High Frontier: Human Colonies in Space*. Garden City: Anchor Books.

———. 1977. "The High Frontier (talk given at 1975 World Future Society Meeting, Washington DC.)." In *Space Colonies*, edited by Stewart Brand, 8–11. Sausalito, CA: Whole Earth Catalog.

———. 1978. "The Low (Profile) Road to Space Manufacturing." *Astronautics and Aeronautics* 16: 18–32.

———. 1981. *2081: A Hopeful View of the Human Future*. New York: Simon and Schuster.

———. 1983. *The Technology Edge: Opportunities for America in World Competition*. New York: Simon and Schuster.

Oravec, Christine. 2000. "An Inventional Archaeology of 'A Fable for Tomorrow.'" In *And No Birds Sing: Rhetorical Analyses of Rachel Carson's Silent

Spring, edited by Craig Waddell, 42–59. Carbondale: Southern Illinois University Press.

Osborn, Fairfield. 1948. *Our Plundered Planet*. London: Faber and Faber.

Page, William. 1973. "Population Forecasting." In *Models of Doom: A Critique of the Limits to Growth*, edited by H. S. D. Cole, Christopher Freeman, Marie Jahoda, and K. L. R. Pavitt, 159–74. London: Chatto & Windus.

Pfaelzer, Jean. 1984. *The Utopian Novel in America, 1886–1896: The Politics of Form*. Pittsburgh: University of Pittsburgh Press.

Pratt, Mary Louise. 1992. *Imperial Eyes: Travel Writing and Transculturation*. London: New York: Routledge.

Presidential Science Advisory Committee. 1963. "Report on Pesticides." John F. Kennedy Presidential Library.

Rojek, Chris. 1997. "Indexing, Dragging and the Social Construction of Tourist Sights." In *Touring Cultures: Transformations of Travel and Theory*, edited by Chris Rojek and John Urry, 52–74. New York: Routledge.

Ruff, Howard J. 1974. *Famine and Survival in America*. 1st ed. s.l.: Ruff.

Said, Edward W. 1993. *Culture and Imperialism*. New York: Vintage.

———. 2003. *Orientalism*. New York: Penguin Books.

Segal, Howard P. 1985. *Technological Utopianism in American Culture*. Chicago: University of Chicago Press.

Shute, Nevil. 1957. *On the Beach*. New York: Ballantine Books.

Smith, John. 1973. *A Map of Virginia with a Description of The Coventrey, The Commodities, People, Government and Religion*. Amster: Theatrum Orbis Terrarum Ltd.

Space Colonies. 1977. Sausalito, CA: Whole Earth Catalog.

Sullivan, Walter. 1974. "Proposal for Human Colonies in Space Is Hailed by Scientists as Feasible Now." *New York Times*, May 13.

Toffler, Alvin. 1970. *Future Shock*. London: Pan.

Vogt, William. 1948. *Road to Survival*. New York: W. Sloane Associates.

Waddell, Craig, ed. 2000. *And No Birds Sing: Rhetorical Analyses of Rachel Carson's Silent Spring*. Carbondale: Southern Illinois University Press.

Westfahl, Gary. 1996. *Islands in the Sky: The Space Station Theme in Science Fiction Literature*. 1st ed. San Bernardino, CA: Borgo Press.

7 Conclusion

A Continuing Mythic Significance

When the space shuttle *Columbia* landed after its maiden flight in April 1981, United Press International (UPI) offered a story that ran nationally across the United States. Under the headline—a quote from the Australian *Sun*'s own coverage—that proclaimed "First Covered Wagon on New American Frontier," UPI's story was full of hope and ran in hundreds of local and regional papers (cf. UPI 1981). But, quickly, that hope was dashed. Dashed with it were O'Neill's utopian dreams of a High Frontier as well as the related frontier dreams of NASA. But myth, of course, can work even when the logos-centered technology of such dreams fails. And searching for William G. Doty's "forms of the future" would continue, if not in reality, then at least in film and fiction.[1] The first space shuttle built for testing, originally to be called *Constitution*, had been renamed *Enterprise* after a letter-writing campaign to President Ford by *Star Trek* fans, after all. And the year after *Columbia*, the first shuttle built specifically for space, was finally launched, the second *Star Trek* film would feature its own *Enterprise* and be released to generally positive reviews (Meyer 1982).

Perhaps surprisingly, even Joe Haldeman would offer frontier-oriented visions in his post–*Forever War* work. That surprise, of course, is that *The Forever War* went to such lengths to repudiate the frontier themes of traditional American science fiction; then Haldeman, at least partially, turned back to such themes as a major shift. As chapter three shows, Haldeman was able to pen an antiwar, anti-Vietnam, antifrontier myths novel in the early 1970s that won every major award of the genre. In returning to such traditional myths, then, Haldeman perhaps offers insight into the continued relevance of those very myths *The Forever War* rejects.

It might, therefore, simply be enough to mention Joe Haldeman's two frontier-themed *Star Trek* novels, written shortly after he published *The Forever War*.

Here was a writer who had offered a major critique of science fictional frontier myths in his first major, award-winning work of the period and who then turned—quickly, in fact—back to the "fold" with such future novels. Written, Haldeman admitted, in part for the money, it is still difficult to see past the frontier-tinged theme of his second *Trek* title, *World without End*

(Haldeman 1977, 1979, 2011). And it could not have been an easy project in that when the show was first being televised, Haldeman was in Vietnam. And, somehow, O'Neill caught Haldeman's attention too. Haldeman's never-finished nonfiction work, which was contracted for publication with New York–based St. Martin's, would be a victim of the shuttle's tardiness. Again, calling on the tropes of his genre and of American myth, it was to have been titled *The Last Horizon* and offer details of a shuttle-based move into orbital space colonies (Haldeman 2011). Additionally, Haldeman's 1981 novel, *Worlds*, begins in such orbiting space colonies with a surprising range of political and social lifestyles much like O'Neill would claim his High Frontier would make possible. One of the first colonies, the novel's narrator explains, was built in 2010 when "smaller-scale energy farms like Devon's World (then called O'Neill) were making fortunes" (2002, 11). Haldeman was clearly inspired by the work of Gerard K. O'Neill, not just in the range of social forms Haldeman presented in his colonies but also in the ways those social lifestyles allow for a subjectively "better" life for the citizens of those colonies.

Of course, Joe Haldeman writes science fiction for a living, and grew up reading the genre. It would be an overstatement to point out that his return to the central themes of that genre is indicative of a return to an American mythological structure, that somehow such themes were inescapable for Haldeman. A far more reasonable suggestion would be that, after writing his Vietnam War novel and after a few years to somewhat overcome the horrendous experiences of the war, Haldeman found himself writing within a genre he knew and cared for. He does not even mention Vietnam in the "author's note" for the *Star Trek* novel *World without End*, stating only, "I was overseas when most of it was aired." Instead, Haldeman moved toward the frontier myths that are the foundations of American science fiction, and O'Neill's High Frontier played a part in that movement. Other writers in the 1980s offered a wealth of new orbiting space colonies, much like those of Haldeman's.

Yet Haldeman was not the only such writer to offer both O'Neill-inspired narratives, as well as other frontier themes. Interestingly, the science fiction writer who had penned the very first *Star Trek* novel in 1967, Mack Reynolds, would also take part in this search for forms of the future. No stranger to the utopian themes espoused by O'Neill, Reynolds had authored "updates" of two of the most famous American utopian novels ever written. His *Looking Backward From the Year 2000* and *Equality in the Year 2000*, written in the 1970s, would rework Edward Bellamy's classic nineteenth-century works (cf. Reynolds [1973, 1977], which rewrite the 1888 and 1897 texts of Bellamy [1967, 1969]). Throughout the decade he would add a number of other utopian-themed "in the year 2000" novels to his oeuvre, all the while collecting royalty checks from his *Star Trek* novel, *Mission to Horatius* (Reynolds 1967; cf. Kapell 2003). But shortly after the publication of O'Neill's *The High Frontier* in 1976, Reynolds also began publishing science fiction novels set in O'Neill's orbital habitats. Like much of Reynolds's work, the prose has not held up well, but his 1983 novel *The Langrangists*, and 1984's *Chaos in Lagrangia*, offer what are essentially spy capers in O'Neill's orbiting space colonies (Reynolds 1983, 1984). *The Langrangists*, just to make sure readers do not miss the point, carries a dedication that says "To Professor Gerard K. O'Neill, Pioneer in the colonization of space." And then, on a wholly different page, an acknowledgment goes one step further as Reynolds makes his debt explicit:

> The present writer would like to extend his appreciation to Professor Gerard K. O'Neill and his staff for permission to quote directly from his various articles in *The Futurist* and scientific publications on the projected Lagrange Five space colony endeavor.[2]

It might not be surprising that Reynolds would publish such works, however. Science fiction writers, as a professional hazard, must adopt current science into their projects and O'Neill represented current popular science at the time. And Reynolds, usually classed as a "social science fiction" writer because of his willingness to deal with sociological and anthropological themes more than those of the hard sciences, was an almost perfect choice for such projects as those inspired by O'Neill and by *Star Trek*.

Between writers like Reynolds and Haldeman, O'Neill space stations moved to the center of American science fiction. Even writers with the very highest of reputations began featuring them. Robert A. Heinlein, whose concept of a magnetic accelerator on the surface of the moon actually figures prominently in O'Neill's plans, began offering O'Neill-themed habitats shortly after the publication of *The High Frontier*. Heinlein was arguably the best-known and most significant American science fiction author after the Second World War, and his works frequently made best-seller lists outside of the genre itself. Although Heinlein was no stranger to the notion of space stations, having offered them in his work for over forty years, in later novels such as *Friday* and *The Cat Who Walks Through Walls*, his stations begin to follow O'Neill's plans and are placed at O'Neill's suggested Lagrange locations (1982, 1985). While the eventual contribution of NASA to the International Space Station has created something much like Heinlein's space stations in his immediate post–World War II fiction, by the early 1980s Heinlein and many other writers would be envisioning the far more expansive O'Neill-inspired space colonies in their fiction. Even the tried-and-true brand of the "final frontier" features O'Neill's plans as, of the first six *Star Trek* films, "every one of them begins or ends with the *Enterprise* docked at, approaching or leaving a space station in Earth orbit," and those stations all bear design features promoted by O'Neill (Westfahl 1996, 64).

Joe Haldeman's *The Forever War*, like much Vietnam-inspired literature, may have suggested that by the end of the 1960s the traditional frontier narratives were hopelessly corrupted. But his later *Star Trek* novels, and his orbiting space colonies novel of *Worlds* argues that those narratives were not as unredeemable as many scholars argued. That other science fiction writers, as divergent as Mack Reynolds and Robert A. Heinlein, were exploring O'Neill's visions is also a strong statement about the continued popularity of the frontier themes not just of O'Neill but also of those themes generally. Similarly, NASA may not have been able to offer the environment their mythic rhetoric called for, and *Star Trek* may remain merely a fictional television show. However, O'Neill's High Frontier, *Star Trek*'s Final Frontier, and NASA's lunar missions suggest that such mythic themes continued to have relevance for sizable groups of Americans. Yet, while such frontiers continued to be featured in many parts of the varied cultures of the United States, scholarship moved further and further away from the possibility of assessing them.

A POSTREVISIONIST MYTHOGRAPHY

Many of the scholarly movements referenced in this work have had revisionist moments. The consensus history of the immediate post–World War II era was seen to readily ignore conflict, and New Left historians and their scholarly descendants reformed the entire discipline of history. The American Studies myth and symbol approach was seen to be too closely aligned with that consensus history and it was cast aside as well. Historians of the frontier finally decided to at least *attempt* to rid themselves once and for all of Frederick Jackson Turner's influence. and they, too, proclaimed a "New" school in the New Western history. And American Studies scholars eventually founded a New Americanist school, which saw *myth* as too static a term and relied, instead, on a pluralistic form of *rhetoric*.

Few scholars would doubt that such "new" schools are unusual as they tend to happen about once a generation in academe. It is interesting to note, however, that movements such as the New Americanists and the New Western history have rejected the study of myth while failing to notice that mythography has had its "new" movements as well. When Philip Fisher, one of the groundswell of New Americanists, noted that "myth is fixed, satisfying, and a stable story used again and again to normalize our account of social life," it is painfully clear he has not read a recent monograph on the study of myth (1991, viii). And when New Western historian Patricia Nelson Limerick declares her movement's fealty to an empirical method and notes this should only be difficult only for those "who have come to depend on illusions" it is similarly conspicuous that she has not kept note of the study of the importance of such "illusions" in the fabric of social life (1991, 86).

"Myths," William G. Doty offers, are "understood as referring to the fundamental religious or philosophical beliefs of a culture, expressed through ritual behavior or through the graphic or literary arts, and forming a constitutive part of a society's worldview" (2000, 13). In its generality, this is not a provocative statement on myth at all and one that even Patricia Nelson Limerick and Philip Fisher could easily accept. But Doty also writes, as has been noted frequently in this work, that "myths . . . anticipate *forms of the future* as they determine and shape ideals and goals for both individual and society" (2000, 29). All the movements found in this work that characterize themselves as "new" would reject such an assertion, simply because it makes their critiques of an analysis of myth impossible to maintain. Many scholarly disciplines worked hard to abandon any "myths of the frontier," because their appraisals considered myth to lack a potential for change. But while such scholars were engaged in the rejection of previous myth-based approaches actual first-class myth scholars were busy noting that myth offered something the critiques claimed it lacked. In refusing to acknowledge that myth offers the possibility of ongoing change academics in various revisionist movements missed the opportunity to find exactly what they desired in their own research. They could have done so without the complete rejection of earlier approaches toward which they had offered, instead, only contempt.

So as Americanists began to look for "dissensus" and found smaller and smaller units of study within the United States, opting for either strict empiricism of history and political science programs or the approved study of rhetoric in various English departments, the study of myth, as a topic of inquiry, had evolved. Additionally, as has been demonstrated in this work, applying the study of myth to a specific period of American history allows for culturally as well as scholarly useful conclusions. The Vietnam War seemed to leave despoiled so much of the mythological structure that had previously offered such utility for the American nation. The reaction to that war strongly implied that those traditional myths would no longer suffice for anticipating Doty's "forms of the future" and that only profoundly "new symbols of possibility," as Marx proclaimed, would now be useful. By writing a science fiction novel that seems to purposefully undermine all the tropes of a genre built upon the traditional frontier myths, Joe Haldeman offered an argument that those myths were no longer valid. *The Forever War* is not unique in this regard. Literature scholars have noted again and again the warped or inverted frontier imagery used by writers about Vietnam. But by offering his "Vietnam novel" in a genre that, in the United States at least, was built on frontier myths, and then undermining those genre traditions, Haldeman offered a serious critique not just of Vietnam, or of war, but also of foundational American myths themselves. And, for a time, much of American society and scholarship agreed with him. Vietnam, it seemed, had revealed a traditional American mythology that was essentially confrontational, imperialistic, racist, and violent. And those few scholars still interested in an American myth of the frontier agreed with that assessment. In venturing only a negative reading of such myths, combined with an ongoing search for "dissensus," such Americanists failed to understand that, in James Oliver Robertson's phrase, "our myths, whether they lead us to positive or negative responses—*and they can do both*—give us a sense that the world is understandable and explicable" (1980, 8; emphasis added). Scholars were not interested in even the *possibility* of positive responses. The political and social disruptions of the late 1960s in the United States, including especially the Vietnam War, had primed them to analyze the negative aspects of traditional myths and conclude, as Vernon E. Mattson did in looking back at myths of the west, that "collective illusions about frontier origins may be dangerous" (1988, 11). Yet in finding nothing positive in such myths—in such "collective illusions"—it was not those specific myths that became worrisome to academics.

It was myths in general.

But the missions of NASA and the starship *Enterprise*, and the "Lagrange Five space colony endeavor" of Gerard K. O'Neill all attempted to offer forms of the future that were understandable and explicable, as well as primarily positive. While NASA still retained too much of a backward-looking mythological system inherited from the Kennedy administration, it was still trafficking in the tropes of the frontier. And those inherited tropes were not just Kennedy's but actually can be traced to the earliest English settlements in North America as well. Unfortunately, they were simply too antiquated to be sufficiently reformed by NASA. Try as the agency might, they would never find on the moon what John Smith found in Virginia, a place that "heaven and earth never agreed better to frame a place for means of habitation." But the agency did try, and twelve men trod the surface of a new world. It might have failed as the first outpost of a new age of discovery, but it remains a significant step both in the logos-centric world of history, technology, and science, as well as in the mythic plane of new and differing frontiers.

Additionally, while *Star Trek* will never be anything other than a multimedia franchise of fictional characters on fictional ships venturing to fictional planets, its popularity cannot be doubted as anything other than the common acceptance of those tropes. It would be simple enough to note the introductory narration of "Space, the Final Frontier," and leave *Star Trek* at that. But, in reality, *Trek* attempted what can only be called a more sophisticated version of frontier myths. Where NASA would work from a many centuries' old script of the New World as a New Jerusalem and a new Garden, *Star Trek* claimed the progress of such discoveries while rejecting those unrealistic Edens. Like NASA's, it remains a frontier myth but one that promulgated a very pragmatic approach to paradise.

Furthermore, while Gerard K. O'Neill's High Frontier appeared destined to failure as soon as the first Space Shuttle launched, its popularity throughout the 1970s also attests to the continuing significance of those myths. While claiming, time and again, he was offering no Utopia, his vision of orbiting "islands in space" came with enough paradisiacal implications that Captain Kirk would certainly have violated the Prime Directive to set O'Neill's colonists on a "normal course of evolution." But the very fact that O'Neill saw nothing unreasonable in his plans and believed they would solve so many earthly troubles attests once again to the powerful agency inherent in these frontier myths. That his plans could not work in the fashion he claimed is unimportant.

That so many wished that they would is.

Of course, these are not just a series of frontier tropes or related types of rhetoric; they each present different iterations of myths with the inherent structure of frontiers. Parts may be fictional, but they are never Limerick's "illusions," and parts may be wishful thinking, but they are never "fixed" or "stable" as Fisher would have them.

They are all cultural experiments in mythmaking.

For a moment in the late 1960s it appeared no such experiments would ever be tried again, or, at the very least, not while bearing any relationship to real or imagined "frontiers." Americanists, in multiple disciplines, and from multiple approaches, could not move beyond the moment when Vietnam appeared to leave American myth bereft of positive imagery, and so they would claim that the American nation could not move beyond that moment, either. When they mentioned myth at all, it was in a negative context. But the many cultures of the United States *were* able to rebuild their traditional myths, with some of them experimenting with those forms in surprisingly detailed new ways.

"I HAVE NEVER BEEN ABLE TO DISCOVER ITS LIMITS"

This work began with the 1839 quote from Caroline Kirkland as she asked of the west, "How much does that expression mean to include? I never have been able to discover its limits." These experiments in mythmaking by NASA, *Star Trek*, and O'Neill, coming as they did after the apparent dissolution of such myths in Vietnam, are also about searching for such limits. In searching for "forms of the future," these experiments in mythology are primed to be culturally significant and are not to be walled in by empirical historiography or reduced to the study of rhetoric. There is no contention here that only these instances are examples of experiments with frontier myths, because of course they are not. Nor is there a contention that the only mythical experiments were those involving ideas of a frontier, for that, too, would be absurd. The question may be asked, as David Adams Leeming has, "there are those who will still ask what all of this has to do with us." Leeming's answer is the answer this book has contended from the first page. It is, Leeming wrote, "that it has everything to do with us" (1981, 5). Myth—or, really, the plural is more appropriate—*myths* are too complex, multifaceted, and polysemous for any declaration that they be limited to what has been explored here. They can be conservative but usually are not. They can look backward but usually offer the view forward just as readily.

Rather than consider myths static, unchanging, and of poor utility for the examination of contemporary culture, this work establishes that an approach using mythography as understood by more recent scholarship is quite appropriate In doing so it provides examples available for analysis that are wide-ranging and significant. Complex patterns of mythmaking within American society can be clearly seen through the examples of NASA, *Star Trek*, and Gerard K. O'Neill. Whether they are empirically or pragmatically successful is not the issue. The simple fact that they were each engaged in mythic undertakings is the most significant part of each of their projects. NASA, *Star Trek* and O'Neill were all attempting to offer what Leo Marx meant in 1964 when he called for "new symbols of possibility." They also erected their myths on existing foundations, and built them with available materials. And in doing so, they offered Doty's "forms of the future." Each was presented with the hope in the sense William James meant about truth, that "it [might] *become* true, is *made* true by events." To argue that those futures never came to be as proof of their failures would be to fundamentally misunderstand the purpose of myths.

Put another way, if Caroline Kirkland never discovered the limits of her "west" is it any surprise that NASA, *Star Trek*, and Gerard K. O'Neill did not, either? This study was not meant to offer an examination of how those narratives found the limits of the west, or of the frontier, in American mythological terms. Instead, this work is an examination of the idea that, beginning in the late 1960s, there were experiments with American mythologies about frontiers. All of them were scouting for Kirkland's "limits," yes, but the purpose of such mythological explorations is not to actually discover boundaries, but to show that they are immaterial. Haldeman thought Vietnam had found those boundaries, but later he partially recanted that position.

That NASA offered to turn the mythos of such frontiers into the logos of the Apollo lunar missions, and in doing so found the limits of a technological frontier does not mean they found the limits of a mythological one. *Star Trek*, too, offered visions of such frontiers, this time televised into American homes, but in its fictional way, it never offered any of the "limits" Kirkland wondered about. For *Trek* was not meant to be about such limits but was instead about proving such limits were mythologically inconsequential. Gerard K. O'Neill's vision of a High Frontier might seem to have found Kirkland's limits with the launching of a mediocre space vehicle in the shuttle, but his real vision was never about such technological impediments but about finding ways to circumvent such practical limitations mythically. So while each vision of frontiers might be said to have failed in some way, such failure only exists on the plane of empirical considerations. From the different plane of a mythological perspective, none of them failed. NASA,

Star Trek, and O'Neill cannot stake a claim to having found anything like Caroline Kirkland's "limits" because that was not their purpose. None of these experiments was actively searching for *limits* at all. Indeed, no other American has ever discovered them, either.

And, in a way, that is *precisely* the point.

NOTES

1 Sources cited in the main chapters of this work are not offered a citation in this conclusion.
2 The quotes appear on unnumbered pages in the front matter of the text.

WORKS CITED

Bellamy, Edward. 1967. *Looking Backward, 2000–1887*. The John Harvard Library. Cambridge: Belknap Press of Harvard University Press.
———. 1969. *Equality*. Grosse Pointe: Scholarly Press.
Doty, William G. 2000. *Mythography: The Study of Myths and Rituals*. 2nd ed. Tuscaloosa: University of Alabama Press.
Fisher, Philip. 1991. "Introduction: The New American Studies." In *The New American Studies: Essays from Representations*, edited by Philip fisher, vii–xxii. Berkeley: University of California Press.
Haldeman, Joe. 1977. *Planet of Judgment*. London: Corgi.
———. 1979. *World Without End*. New York: Bantam.
———. 2002. *Worlds*. London: Gollancz.
———. 2011. "Email on Star Trek, Vietnam, and The Forever War, to Matthew Wilhelm Kapell," May 6.
Heinlein, Robert A. 1982. *Friday*. New York: Holt, Rinehart and Winston.
———. 1985. *The Cat Who Walks Through Walls: A Comedy of Manners*. New York: Putnam.
Kapell, Matthew. 2003. "Mack Reynolds' Avoidance of His Own Eighteenth Brumaire: A Note of Caution for Would-Be Utopians." *Extrapolation* 44 (2): 201–8.
Leeming, David Adams. 1981. *Mythology: The Voyage of the Hero*. 2nd ed. New York: HarperCollins.
Limerick, Patricia Nelson. 1991. "What on Earth Is the New Western History?" In *Trails: Toward a New Western History*, edited by Limerick, Patricia Nelson, Clyde A. Milner, and Rankin, Charles E., 81–8. Lawrence: University of Kansas Press.
Mattson, Vernon E. 1988. "West as Myth." *Journal of the History of the Behavioral Sciences* 24 (1): 9–12.
Meyer, Nicholas. 1982. *Star Trek II: The Wrath of Khan*. Los Angeles, CA: Paramount Pictures.
Reynolds, Mack. 1967. *Mission to Horatius*. New York: Pocket Books.
———. 1973. *Looking Backward from the Year 2000*. New York: Ace.
———. 1977. *Equality in the Year 2000*. New York: Ace.
———. 1983. *The Lagrangists*. New York: Tor.
———. 1984. *Chaos in Lagrangia*. New York: Tor.

Robertson, James Oliver. 1980. *American Myth, American Reality*. New York: Hill & Wang.

United Press International. 1981. " 'First Covered Wagon on New American Frontier': US. Space Success Hailed Worldwide." *Daily Record*, April 15, sec. A.

Westfahl, Gary. 1996. *Islands in the Sky: The Space Station Theme in Science Fiction Literature*. 1st ed. San Bernardino, CA: Borgo Press.

Index

www.ingramcontent.com/pod-product-compliance
Ingram Content Group UK Ltd.
Pitfield, Milton Keynes, MK11 3LW, UK
UKHW022000180926
475701UK00003B/116
* 9 7 8 0 3 6 7 2 6 3 9 6 6 *